Jörg Schänzlin

Modeling the long-term behavior of structural timber

Jörg Schänzlin

Modeling the long-term behavior of structural timber

Modeling the long-term behavior of structural timber for typical serviceclass-II- conditions in South-West Germany

Südwestdeutscher Verlag für Hochschulschriften

Imprint
Any brand names and product names mentioned in this book are subject to trademark, brand or patent protection and are trademarks or registered trademarks of their respective holders. The use of brand names, product names, common names, trade names, product descriptions etc. even without a particular marking in this work is in no way to be construed to mean that such names may be regarded as unrestricted in respect of trademark and brand protection legislation and could thus be used by anyone.

Publisher:
Südwestdeutscher Verlag für Hochschulschriften
is a trademark of
Dodo Books Indian Ocean Ltd., member of the OmniScriptum S.R.L Publishing group
str. A.Russo 15, of. 61, Chisinau-2068, Republic of Moldova Europe
Printed at: see last page
ISBN: 978-3-8381-2462-9

Zugl. / Approved by: Stuttgart, Universität Stuttgart, Habilitation, 2010

Copyright © Jörg Schänzlin
Copyright © 2011 Dodo Books Indian Ocean Ltd., member of the OmniScriptum S.R.L Publishing group

Acknowledgment

This study has been carried out at the Institute of Structural Design at Universität Stuttgart. Hereby I want to thank Prof. Dr.-Ing. Ulrike Kuhlmann, Institute of Structural Design, Universität Stuttgart, for the possibility and the supervision of this study, for the support during my whole time at the University and the belief in the feasibility of this study.
Additionally, I want to thank Prof. Sven Thelandersson, PhD, Lunds Universitet, Sweden for the possibility to start this study at his institute and for the co-supervision of this study. Beside that, I also want to thank Prof. Dr.-Ing. Klaus Sedlbauer, Universität Stuttgart, for the co-supervision, the fruitful discussions and the comments to this study.

I want to thank Dr. Simon Aicher, MPA Stuttgart, Dipl.-Ing. Matthias Gerold, Harrer Ingenieure, Karlsruhe, Assoc.-Prof. Massimo Fragiacomo, University of Sassari, Italy, Assoc.-Prof. Luis Jorge, Polytechnic Institute of Castelo Branco, Portugal and Assoc.-Prof. Dr. Alfredo Geraldes Dias, University of Coimbra, Portugal for their support and their interest in my work during the whole time at Universität Stuttgart.

Besides that, I want to thank Prof. Dr.-Ing. Wolfgang Francke and the former students Dipl.-Ing. Falko Dieringer, Andreas Lühr M.E. and Marco Gutenkunst B.E. of the HTWG Konstanz, who had decided to examine creep coefficients in real structures in their final theses, thereby contributing directly to this study.

I also want to thank Gerd Futter, Ofterdingen and Rainer Klett, Universitätsstadt Tübingen, Fachabteilung Hochbau for their support and the opportunity to perform the measurements in their buildings even though the use of the buildings was limited during the measurements.

Additionally, I want to thank the members of the Institute especially Dipl.-Ing. Gabriele Teichmann, Ing. Pietro Aldi, Dipl.-Ing. Frank Brühl and Dipl.-Ing. Reiner Hofmann for their support, their questions, their remarks to this and all the other topics and for the pleasant atmosphere at work.

Finally I want to thank my whole family, especially Antje and our little son Finn Lukas, who enabled me to think about the unsolved questions even at times, when I normally used to sleep deeply.

Nomenclature

c	concentration
D	diffusion coefficient
E	Modulus of Elasticity
F	external single load
g	permanent load
h	relative vapor pressure
J	compliance of the model
k_{def}	creep coefficient according to the standards
K	stiffness of the spring
L	length or span of an element
M	bending moment
MC	moisture content
N	normal force
RH	relative humidity
S	surface emissivity
SCL	service class according to the standards
t	time
u	moisture content
w	deformation/deflection
q	external maximum load (= dead load + live load)
α	shrinkage/swelling coefficient
ΔRH	amplitude of the annual relative humidity
ε	strain
κ	curvature
σ	stress
τ	retardation time of the model
φ	evaluated or measured creep coefficient

Subscripts:

cr	creep
el	elastic
eq	equilibrium
fin	final
ms	mechano-sorptive
0	initial
tot	total
∞	end of the regarded periods (=50 years)

Definitions

creep	increase of the strain for a constant stress
normal creep	increase of the strain for a constant stress due to the duration of load
mechano-sorptive creep	increase of the strain for a constant stress due to moisture variations
relaxation	decrease of the stress for a constant total strain

Contents

Acknowledgment	1
Nomenclature	3
1 Introduction	**1**
2 State of the art	**3**
2.1 Motivation	3
2.2 Influences on the creep strain	6
2.3 Explanatory models for the creep behavior	7
2.4 Considering creep in the structural design	10
3 Modeling of the long-term behavior of timber	**13**
3.1 Tool *kriHo*	13
3.2 Rheological models	16
3.2.1 General	16
3.2.2 Model according to Toratti [51]	16
3.2.3 Model according to Hanhijärvi [22]	19
3.2.4 Model according to Becker [2]	20
3.2.5 Model according to Mårtensson [38]	24
3.3 Summary	30
4 Comparison of the creep coefficients	**31**
4.1 General comparison of the models	31
4.2 Constant surrounding conditions	32
4.3 Effective creep coefficients in variable climate	35
4.4 Influence of the variability of the material properties on the creep coefficient	37
4.5 Effects for the application in timber-concrete composite structures	38
4.6 Conclusions	39
5 Determination of the existing long-term deformations	**41**
5.1 General	41
5.2 Determination of creep coefficient by measurements	42

	5.2.1	Procedure	42
	5.2.2	How to deal with the unknown influences?	43
5.3		Determination of effective creep coefficients in the region of Tübingen, South-West Germany	51
5.4		Comparison of the measured creep coefficients to the responses of the models, to common rules in the standards and to studies from literature	55
5.5		Summary	58

6 Modified rheological model for the re-evaluation of the creep coefficients 59
- 6.1 General . . . 59
- 6.2 Modification of model B according to Toratti [51] . . . 64
- 6.3 Comparison to measurements in the region of Breisgau-Hochschwarzwald . . 65
- 6.4 Conclusions . . . 69

7 Simplified equations for the determination of creep deformations 71
- 7.1 General . . . 71
- 7.2 Global creep effects based on model B according to Toratti [51] and the modified model B . . . 73
  - 7.2.1 Creep coefficients for constant environmental conditions . . . 73
  - 7.2.2 Consideration of the changing moisture while reaching the equilibrium moisture content after 50 years . . . 74
  - 7.2.3 Influence of the moisture content, changing in one-year sinusoidal cycles 77
    - 7.2.3.1 Determination of the minimum moisture variation within the cross section . . . 77
    - 7.2.3.2 Determination of creep coefficients considering the global mechano-sorptive creep . . . 80
  - 7.2.4 Interaction between drying and changing moisture . . . 81
- 7.3 Local creep effects . . . 82
  - 7.3.1 General . . . 82
  - 7.3.2 Determination of effective dimensions for the model according to Toratti [51] . . . 83

8 Influence of the load history 85
- 8.1 General . . . 85
- 8.2 Influence of the duration of load in constant climate . . . 86
- 8.3 Influence of an annually repeating load, e.g. snow in variable climate . . . 90
- 8.4 Comparison to the measurements by Gutenkunst [20] . . . 94

9 Effects of the increased creep deformations on the behavior of systems 97
- 9.1 Influences on the ultimate resistance of columns . . . 97

	9.1.1 General	97
	9.1.2 Determination of the critical ratio of the permanent load and the total load based on DIN 1052 [9] and based on the modified creep coefficient	97
9.2	Influence on timber-concrete composite slabs	101
9.3	Conclusions	104

10 Conclusion and Outlook　105

11 Bibliography　109

A Verification of the implementation　113
- A.1 General . . . 113
- A.2 Model B according to Toratti [51] . . . 113
- A.3 Model according to Hanhijärvi [22] . . . 114
- A.4 Model according to Becker [2] . . . 115
- A.5 Model according to Mårtensson [38] . . . 117

B Bionic approach YaRM 0.1　119
- B.1 General . . . 119
- B.2 Modification of the Boyd [4]'s explanation model for the evaluation of the creep strains . . . 120
- B.3 Determination of the elastic properties . . . 121
- B.4 Shrinkage coefficient of the gel . . . 128
- B.5 Consideration of the moisture content on the elastic properties . . . 128
- B.6 Consideration of creep deformations in the model . . . 129
  - B.6.1 General . . . 129
  - B.6.2 Determination of the creep parameters . . . 130
    - B.6.2.1 Basic creep function . . . 130
    - B.6.2.2 Creep parameter based on the creep behavior parallel and perpendicular to grain . . . 131
    - B.6.2.3 Determination of the creep parameters on a simplified model 132
- B.7 Comparison of the bionic approach YaRM 0.1 to test results from literature . 135
  - B.7.1 Constant climate . . . 135
  - B.7.2 Variable climate . . . 136
- B.8 Conclusions . . . 138

C Determination of creep coefficients according to the rheological models　141
- C.1 General . . . 141
- C.2 Global effects . . . 141

- C.2.1 Creep coefficients according to Hanhijärvi [22] 141
  - C.2.1.1 Creep coefficients for constant environmental conditions . . 141
  - C.2.1.2 Consideration of the changing moisture while reaching the equilibrium moisture content after 50 years 142
  - C.2.1.3 Influence of the moisture content, changing in one-year sinusoidal cycles 143
  - C.2.1.4 Interaction between drying and changing moisture 145
- C.2.2 Creep coefficients according to Becker [2] 145
  - C.2.2.1 Creep coefficients for constant environmental conditions . . 145
  - C.2.2.2 Consideration of the changing moisture while reaching the equilibrium moisture content after 50 years 146
  - C.2.2.3 Influence of the moisture content, changing in one-year sinusoidal cycles 146
  - C.2.2.4 Interaction between drying and changing moisture content . 147
- C.2.3 Creep coefficients according to Mårtensson [38] 148
  - C.2.3.1 Creep coefficients for constant environmental conditions . . 148
  - C.2.3.2 Consideration of the changing moisture while reaching the equilibrium moisture content after 50 years 149
  - C.2.3.3 Influence of the moisture content, changing in one-year sinusoidal cycles 149
  - C.2.3.4 Interaction between drying and changing moisture 151
- C.3 Local effects 151
  - C.3.1 Determination of effective dimensions for the model according to Hanhijärvi [22] 151
  - C.3.2 Determination of effective dimensions for the model according to Becker [2] 152
  - C.3.3 Determination of effective dimensions for the model according to Mårtensson [38] 153

D Differences in the modeling of the time-dependent behavior due to the used rheological bodies   155
- D.1 General 155
- D.2 Basic differential equation of parallel Maxwell-bodies 155
- D.3 Basic differential equation of serial Kelvin-Voigt-bodies 156
- D.4 Equality of both principal models 157

1 Introduction

When timber is loaded over a certain time, the deformation of the structural element increases. This increase of the deformation influences the serviceability as well as the ultimate limit state. For the design creep coefficients are given in the different standards. However, quite different temporal deformations are expected, when comparing these creep coefficients to each other. One reason for these differences is, that not all influences on the creep behavior are considered in the different standards (see Chap. 2).

In order to evaluate the temporal deformation of timber with respect to the most important influences several rheological models have been developed by various researchers (see Chap. 3).

In Chap. 4 some of these rheological models are compared to each other. As a result quantitative as well as qualitative differences in the responses of the models are evaluated. A major cause for the differences between the models is, that — strictly spoken — they can only be used for the period of time, for which they have been validated by comparing to test results. Within this period ($\approx$ 5 years) the differences between the models seem to be negligible for the practical use in the structural design. However, extrapolating the models to a period of time of 50 years, significant differences between the responses of the models become apparent. So the question arises, which model should be used for the prediction of the time-dependent behavior of timber.

In order to choose the model, that fits the "real" time-dependent deformations best, the deformations of structural elements in buildings in the region of Tübingen, South-West Germany, have been measured (see Chap. 5). Since the existing deformation is known and the elastic global stiffness of the element can be determined by a defined loading, the effective creep coefficients can be evaluated. These effective creep coefficients represent the creep coefficients which should have been used by the engineer in order to predict the "real" deformation after 50 years including as a still indistinct aspect the imperfections, which influence the result.

Comparing the effective creep coefficients to the responses of the various models, differences between the models and the measured creep coefficients can be found. Therefore the model with the lowest differences between the measured and the evaluated creep coefficients is modified in order to correspond to the measurements (see Chap. 6). As an additional verification, measurements are also performed in the region of Breisgau-Hochschwarzwald. These results are also compared to the responses of the modified model. This comparison leads to a sufficient correspondence between the measurements and the creep coefficients evaluated by the modified model. For a simplified determination of the time-dependent deformation equations are adapted to the numerically evaluated creep coefficients (see Chap. 7).

In experiments, the load is often constant over time. However, in real structures, the external load varies. For example, the elements of roof structures are permanently loaded by the dead load, whereas the snow load acts only during the winter period. This non-permanent load may also lead to creep deformations. For practical use, the effect of this temporal load is considered by splitting this temporal load into a quasi-permanent load and a short-term load. The quasi-permanent load leads to creep deformation, whereas the short-term loads do not create any creep deformation. In Chap. 8 different scenarios are evaluated, so the ratio ψ between the quasi-permanent live load and the total live load can be determined.

As mentioned before the creep deformations do not only influence the serviceability limit state, but also the ultimate limit state, especially when dealing with hyperstatic systems or when elements are subjected to compression. In Chap. 9 the effects of the measured creep coefficients on the structural performance of columns are studied as well as the consequences of the increased creep coefficients on the internal forces and deformations of timber-concrete composite slabs and beams. Following the results of this study, creep deformations in columns have to be considered for lower ratio between the permanent and the total load than given in the German standard DIN 1052 [9]. Concerning the evaluation of the stresses and the effective creep deformations in timber-concrete-composite slabs, the intervals for the evaluation of the effective creep coefficients according to [48] have to be modified, leading to larger deformations and larger stress redistributions in time compared to the procedure given in [48].

2 State of the art

2.1 Motivation

In order to use buildings over their whole life time, the load capacity as well as the serviceability have to be ensured during the whole period of time. During this period, the structural elements have to carry permanent as well as short-term loads. Especially the permanent loads – such as dead loads – lead to an increase of the deformation (= creep deformation) and to a reduction of the resistance of the timber elements.

The long-term behavior of timber elements can be divided into three different phases (see Kenel and Meierhofer [31] and Fig. 2.1)

- primary creep phase: During this phase the deformation increases rapidly until the increment of the deformation becomes stable.
- secondary creep phase: During this phase a constant increment of the creep deformation appears.
- tertian creep phase: The deformation increases disproportionately, which finally leads to a failure of the element.

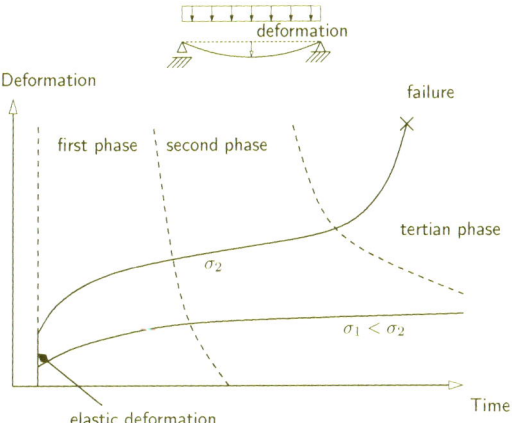

Figure 2.1: Creep phases

Whether timber elements reach the tertian creep phase within the lifetime of the building depends on the stress level. Therefore the resistance is modified in dependence on time e.g. in the standards DIN 1052 [9] and Eurocode 5 [14]. Due to the reduction of the resistance, no creep failure should occur in a building during its lifetime, so the tertian creep phase should not be reached. Since the tertian creep phase is not reached, only the creep deformations of the second phase have to be considered for determining the deformations and the internal forces within the structural design.

For the serviceability limit state, the increase of the deformation due to the time-dependent behavior is considered by means of creep coefficients. In the ultimate limit state, in general the time-dependent behavior is only considered in the modification of the resistance of the material. However in some situations creep deformations can influence the internal forces and therefore the stresses, especially in the following situations:

- stability of columns or lateral torsional buckling: The load capacity of these elements depends on the imperfection of the system. Creep deformations can be interpreted as an additional imperfection of the system at a certain point in time (see Fig. 2.2 and Hartnack [24]). Therefore the load capacity of the system depends on the creep deformation of the system

$$w_{0,eff} = w_0 + w_{cr} = w_0 + k_{def} \cdot w_{el} \tag{2.1}$$

where $w_{0,eff}$ effective imperfection of the column
w_0 imperfection at time $t = 0$
w_{cr} creep deformation
w_{el} elastic deformation
k_{def} creep coefficient

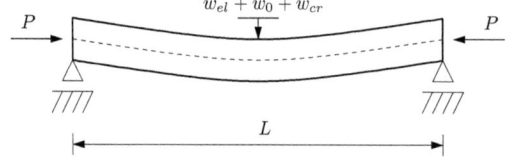

Figure 2.2: Imperfection of a column

- hyperstatic systems and composite structures: In composite structures, as e.g. in

Figure 2.3: Timber-concrete composite elements

2.1 Motivation

timber-concrete composite systems (see Fig. 2.3), the stress distribution within the composite system depends on the stiffness of the single elements. Since creep can be interpreted as a reduction of the long-term stiffness as given in Eq. (2.2), it directly influences the internal stress distribution.

$$E_{eff} = \frac{E_M}{1+\varphi} \qquad (2.2)$$

where E_{eff} effective Modulus of Elasticity
        E_M Modulus of Elasticity of the material
        φ effective creep coefficient in the composite system (see among others [47–49])

- user modifications to "re-ensure" the serviceability: Creeping of timber leads to larger deflections. These larger deflections sometimes limit the use of the building in its intentional sense. For example, the time-dependent deformation of the beam of the

Figure 2.4: Shed in the forest of Tübingen

shed in Fig. 2.4 has increased more strongly than initially expected. Due to the increased deformation the door could not be opened anymore. The options were to cut the door at the top, so it would fit underneath the eaves beam, or to notch the eaves beam. In this case the owner decided to notch the beam, since he probably did not want to have a gap between the door and the door frame. At first glance, everything looks fine:

  – the door can be opened again
  – the interior is protected, since no hole in the wall with door closed.

However, the load capacity of the beam is reduced. The original problem of fulfilling the requirements of serviceability turns into a problem in the ultimate limit state due to the user modifications. Besides that, it is expected that the deformation of the beam will get stronger.

As shown in the three examples above, creep deformations do not only influence the serviceability limit state, but may also influence the ultimate limit state.

2.2 Influences on the creep strain

The creep strain of timber depends on several parameters (see a. o. Moorkamp [40]). The most important influences are

- stress level: Gressel [17] studied the influence of different load level (see Fig. 2.5), concluding that the creep coefficient increases with increasing load level.

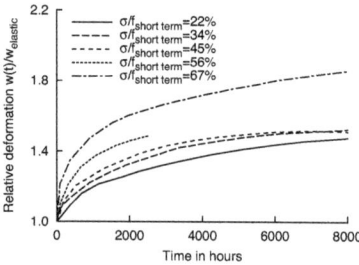

Figure 2.5: Creep coefficients in dependence on the stress level (see Gressel [17])

- type of loading: In Gressel [17] typical creep coefficients of timber subjected to tension, compression, bending and shear are shown (see Fig. 2.6). As can be seen, the creep

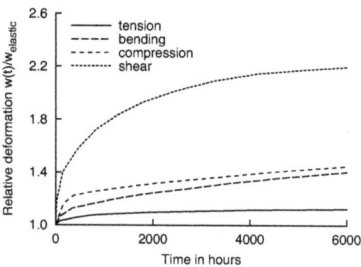

Figure 2.6: Typical creep coefficients in dependence on the type of loading (see Gressel [17])

due to shear reaches the maximum creep deformation, whereas tension leads to the minimum creep deformation.

- moisture content and moisture variation: In Fig. 2.7 an example of the creep coefficient of a timber element is shown, which is subjected to changing relative humidity of the surroundings given in Fig. 2.7(a). The creep deformation increases with every moisture variation. The creep caused by the variation of the moisture content is called mechano-sorptive creep.

- temperature: Morlier and Palka [41] studied the influence of the temperature on the creep coefficient (see Fig. 2.8). Fig. 2.8 shows, that there is an influence of the temperature; however, in the normal ranges of temperature up to $\approx 50°C$ this influence can be neglected.

2.3 Explanatory models for the creep behavior

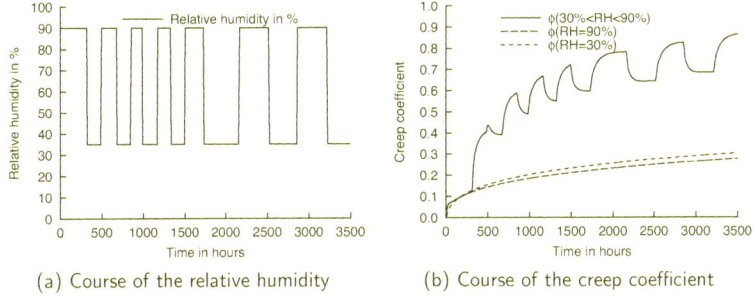

(a) Course of the relative humidity

(b) Course of the creep coefficient

Figure 2.7: Increase of the deflection due to changing moisture content (see Hanhijärvi [22])

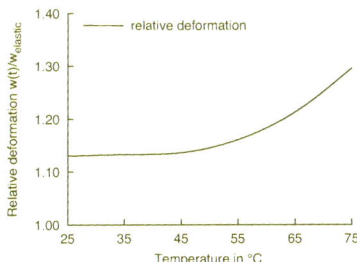

Figure 2.8: Influence of the temperature on the creep deformation after one week (see Morlier and Palka [41])

- force-to-grain-angle: Among others, Niemz [44] studied the influence of the force-to-grain-angle, concluding that the creep coefficient perpendicular to grain is about 8 times larger than the creep coefficients parallel to the grain.

2.3 Explanatory models for the creep behavior

For the explanation of the creep behavior several models exist[1]. Three of them are presented in the following:

- creep as a reaction of the microfibrils (see Boyd [4]): According to Boyd [4] the fibres of timber consist of microfibrils, in between a viscous gel is embedded. Therefore the normal creep strain (=creep strain dependent on time) is caused by the yielding gel. The mechano-sorptive creep (=creep strain dependent on moisture variations) is caused by shrinkage of the gel, resulting in a gap between the microfibrils and the gel. Loading the microfibrils in this situation results in an increased deformation (see Fig. 2.9).

---
[1] Moorkamp [40] and Hanhijärvi [22] give a good summary about the different explanations. So most of this section is taken from these sources.

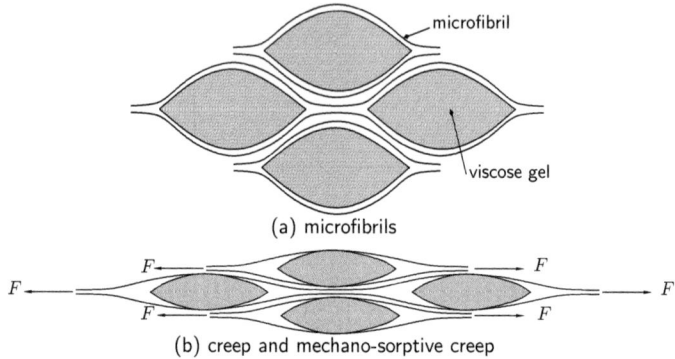

Figure 2.9: Boyd [4]'s model for the explanation of creep and mechano-sorptive creep

- creep and mechano-sorptive creep as a result of the breaking of hydrogen bonds (see Grossmann [19]): According to Grossmann [19] creep is caused by the breaking of the hydrogen bonds between the cellulose chains in the timber. Stronger loaded hydrogen bonds will break, resulting in a slip between the single cellulose chains. So the material is creeping. Since the mechano-sorptive creep is caused by a variation of the moisture content, it is assumed, that the water penetrates into the hydrogen bonds between the cellulose chains. Therefore some hydrogen bonds break, resulting in an additional slip between the cellulose chains (see Fig. 2.10) and in a reduction of the Modulus of Elasticity. However, hydrogen bonds can be formed again in the new configuration.

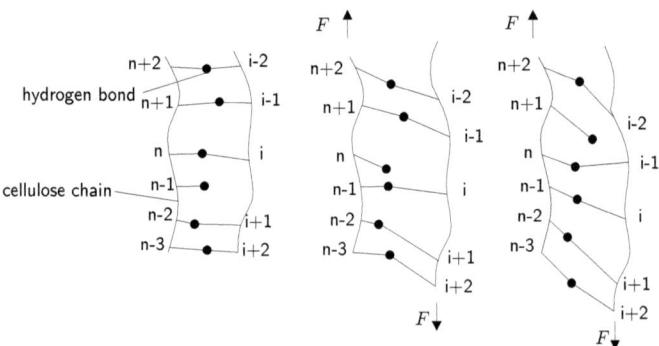

Figure 2.10: Creep due to the movement of the hydrogen bond (see Grossmann [19] and Hanhijärvi [22])

- slip-plane-model for the explanation of creep in compression (see Hoffmeyer and Davidson [25]): Hoffmeyer and Davidson [25] explain the mechano-sorptive creep in compression by the formation of slip planes. These slip planes can be interpreted as a local

2.3 Explanatory models for the creep behavior

buckling of the microfibrils. Since the stiffness of timber in general depends on the moisture content, the resistance against the local buckling decreases with increasing moisture content, resulting in a larger amount of slip planes (see Fig. 2.11).

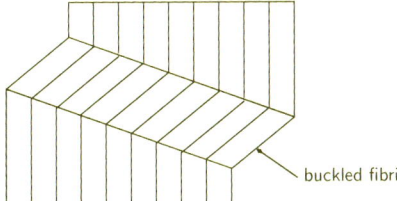

Figure 2.11: Creep as buckling of the fibrils (see Hoffmeyer and Davidson [25] and Hanhijärvi [22])

By these models, most of the influences on the long-term behavior of timber can be explained.
- influence of the stress level: The higher the load in the cellulose chains the more hydrogen bonds will break, resulting in a larger slip between the cellulose chains. However, the larger the slip, the lower the probability that broken parts of hydrogen-bonds will find their partners in order to remake the connection between the cellulose chains. Therefore the amount of bonds is reduced, resulting in a reduced load capacity and a softer system.
- influence of the type of loading: In the explanation model according to Boyd [4] the cell wall of timber consists of microfibrils, in between a viscose gel is included. In dependence on the type of loading (see Fig. 2.12), the creep deformations differ,

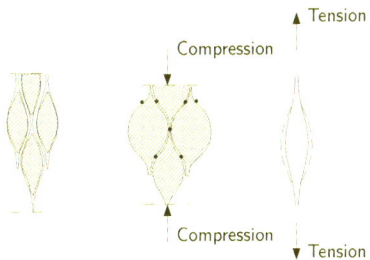

Figure 2.12: Reaction of the cell wall during loading (see Boyd [4] and Hanhijärvi [22])

because under tension the microfibrils are stretched and therefore the loading of the gel is reduced. Thus the creep coefficient under tension is lower than the creep deformation under compression.
- absolute moisture content and moisture variation: The influence of the level of the moisture content can be explained by the strong dependence of the gel on the moisture content in the model according to Boyd [4]. Since increasing moisture reduces the viscosity of the gel, the creep deformation increases (see Fig. 2.7).

According to Grossmann [19], creep is caused by the breaking of hydrogen bonds. In order to consider the mechano-sorptive creep, Grossmann [19] introduces strong and weak hydrogen bonds. The weak bonds can be broken by water, so the cellulose chains can move relating to each other during moisture variation.

- temperature: If the properties of the gel depend on the temperature, the creep behavior of the whole system "microfibril with interlayered gel" is influenced by the temperature.
- force-to-grain-angle: Loading the microfibrils perpendicular to the grain, the gel in between the microfibrils is loaded directly. Since the properties of the gel differ from the properties of the microfibrils, the creep behavior perpendicular to grain differs from the behavior parallel to grain.

2.4 Considering creep in the structural design

In order to consider the time-dependent behavior in the structural design of timber elements, creep coefficients are defined by the standards (see DIN 1052 [8], DIN 1052 [9], DIN 1074 [11] and Eurocode 2 [13]). These creep coefficients describe the ratio between creep strain and elastic strain.

$$k_{def} = \frac{\varepsilon_{cr}}{\varepsilon_{el}} \qquad (2.3)$$

where ε_{cr} strain due to creep
       ε_{el} elastic strain
       k_{def} creep coefficient

A comparison between the creep coefficients according to the standards Eurocode 5 [14], DIN 1052 [8], DIN 1074 [11] and according to the studies by Gressel [17] shows different results of the creep coefficient (see Fig. 2.13 and Fig. 2.14), although the creep coefficients in Fig. 2.13 and Fig. 2.14 should describe the same behavior and therefore should lead to the same result.

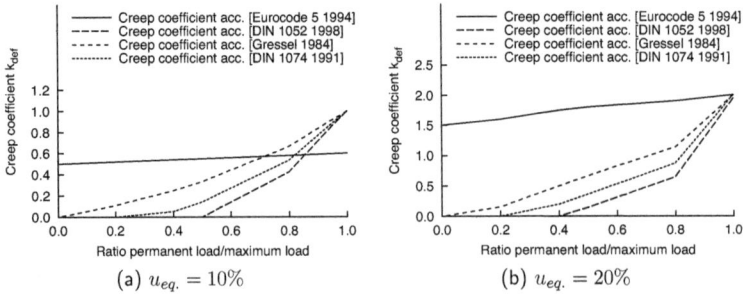

(a) $u_{eq.} = 10\%$

(b) $u_{eq.} = 20\%$

Figure 2.13: Creep coefficient for solid timber in dependence on the ratio of permanent loading g to the total load p, assuming, that the live load does not affect any creep

2.4 Considering creep in the structural design

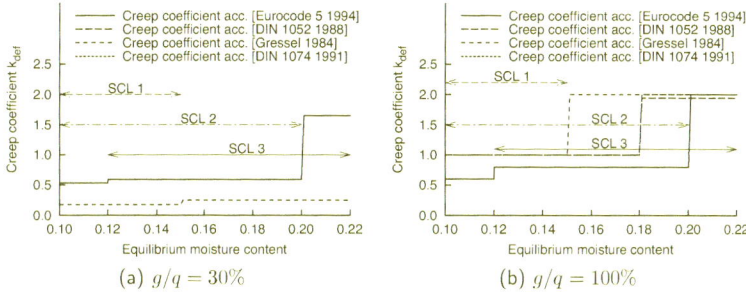

Figure 2.14: Creep coefficient for solid timber in dependence on the equilibrium moisture content $u_{eq.}$ (SCL = service class)

Table 2.1: Range of the creep coefficient of timber k_{def} according to different standards Eurocode 5 [14], DIN 1052 [8] and DIN 1074 [11]

	k_{def}	$(k_{def,max} - k_{def,min})/k_{def,max}$
equilibrium moisture content of 10%		
g/q=1.0	0.6 ... 1.0	40%
g/q=0.6	0.1 ... 0.6	83%
g/q=0.3	0.0 ... 0.5	100%
equilibrium moisture content of 15%		
g/q=1.0	0.8 ... 2.0	60%
g/q=0.6	0.1 ... 0.7	85%
g/q=0.3	0.0 ... 0.6	100%
equilibrium moisture content of >20%		
g/q=1.0	1.9 ... 2.0	5%
g/q=0.6	0.1 ... 1.8	95%
g/q=0.3	0.0 ... 1.7	100%

However, clear differences between the various standards arise (see Tab. 2.1). The differences between Eurocode 5 [14] and the other standards cannot be explained by the different safety levels, because the creep coefficients aim at the determination of the deflection. The deflection is determined in Eurocode 5 [14] as well as in the other standards mentioned, based on the characteristic loads and average material properties.

As a result of the different creep coefficients, in some cases, especially for small permanent loadings ($g/q < 0.5$), the slenderness of the cross-sections can be restricted by the limitation of the maximum deflection in the design according to Eurocode 5 [14], although the creep deformation need not be considered in the structural design according to DIN 1052 [8] in these cases (see Fig. 2.15).

Due to the non-uniform creep coefficients of the different standards (see Tab. 2.1) the question arises, which creep deformation can be expected in reality and whether the creep

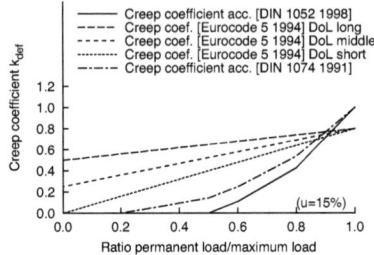

Figure 2.15: Creep coefficient of solid timber in dependence on the duration of load (DOL), assuming that the complete live load consists of loads of only one class of duration of loads

deformation can be determined in a sufficiently exact way.

One reason for the different creep coefficients is the consideration of different influences. So two concepts are generally used in the standards:

- creep coefficients based on the stress level of the cross section due to permanent load, as in DIN 1052 [8] or DIN 1074 [11]
- creep coefficients based on the surrounding conditions as in DIN 1052 [9] and Eurocode 5 [14]; the different duration of load is considered by a quasi-permanent part of the live load

The determination of the creep coefficients is therefore based on different approaches, because the duration of load does not depend on the stress level and vice versa. Both concepts consider the average moisture content. However, the consideration of the equilibrium moisture content is evaluated in steps. Despite the different creep coefficients, the long-term behavior of timber cannot be considered completely by the creep coefficients of the different standards, since not all influences mentioned in Sec. 2.2 are taken into account.

Modelling the time dependent behaviour according to the standards leads to quite different expected deformations after a period of 50 years. In order to give more precise results for the deformation, several researchers have developed various models in order to consider these missing influences. For the systematical determination of the effects of these influences and a desired simplified determination of the long-term behavior of timber, the various responses of these models under different surrounding conditions are of interest. A systematical comparison of these models and a determination of the effective creep coefficients have not yet been performed. For this reason in the following the models according to Toratti [51], Hanhijärvi [22], Becker [2] and Mårtensson [38] are used in order to discuss the effects of the various influences on the effective creep coefficients of structural timber.

3 Modeling of the long-term behavior of timber

3.1 Tool *kriHo*

A comparison of studies concerning the time-dependent behavior of timber-concrete composite slabs (see a. o. Fragiacomo [15], [48], Grosse et al. [18], Bou Saïd [3]) lead to quite different results of the deformations and the internal forces of the single models for the composite structures. One difference between the models mentioned above is the used rheological model of timber. So the question arises, whether the different modeling of the time-dependent behavior is the main reason for these quite different results in the evaluation of the time dependent behavior of timber-concrete composite structures.

In order to compare various rheological models of timber, a C++-tool named *kriHo* has been developed in a modular manner to the effect, that various rheological models can be linked into the tool. This model provides all necessary functions, such as determination of the internal forces, of the deflection, of the moisture content, etc. except for the functions dealing with the stress-strain relation and the coefficients of the moisture transport within the cross section (see Fig. 3.1). These functions are provided by the rheological model.

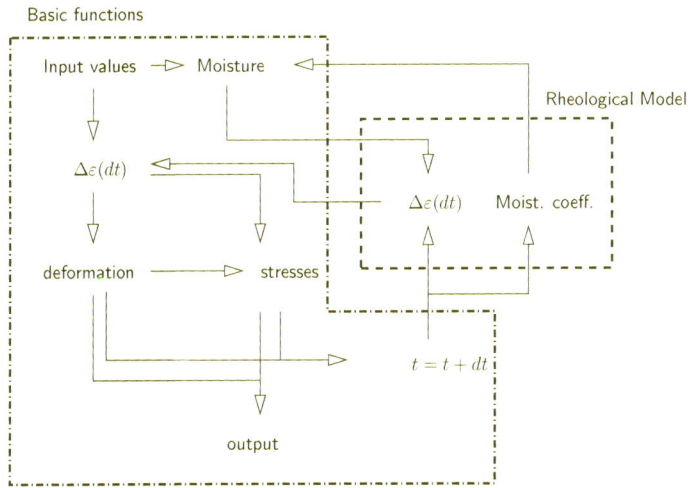

Figure 3.1: Calculation procedure in *kriHo*

The temporal deflection, strains and stresses are determined by following steps:
- evaluation of the increase of the internal forces and stresses respectively due to external loads within a time step Δt

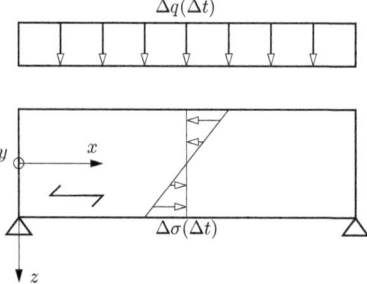

- evaluation of the current moisture distribution within the cross section, assuming that the moisture distribution is independent of the span direction.

$$\frac{du}{dx} = 0 \tag{3.1}$$

- determination of the increase of the strains due to shrinkage/swelling and creep assuming, that the system is fixed in every node

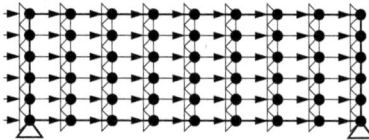

- statical reduction of the system, assuming that Bernoulli's hypothesis is valid

$$\varepsilon(z) = \varepsilon_{\text{centroid of MoE}} + \kappa \cdot z \tag{3.2}$$

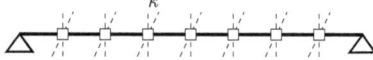

- applying of the internal forces in the fixed system

- determination of the elastic deformation
- determination of the stresses

The problem is transformed from creep to relaxation within the time step Δt. As studies show, the influence of the length of the time step is negligible if the time step chosen is less than 6 hours. Therefore the change of the stresses within the time step Δt need not be considered.

By adding all different rheological models to the tool, it is ensured that all models are used within the same implementation, so no differences due to different formulations of the solving algorithm will occur.

3.1 Tool kriHo

In the rheological models the moisture content and the variation of the moisture content in each point of the cross section influence the time-dependent strain. Therefore the moisture content has to be known. It is common practice to use Fick's law and to neglect influences caused by the anisotropy of timber. Based on this assumption the moisture flow q can be determined in dependence on the water concentration c and the diffusion tensor $\mathbf{D}$ (comp. among others Hanhijärvi [22])

$$\mathbf{q} = -\mathbf{D} \cdot \nabla c \tag{3.3}$$

As second relation the conservation law is valid.

$$\frac{\partial c}{\partial t} = -\nabla \mathbf{q} \tag{3.4}$$

Combining these equations with the relation of the moisture content u and the concentration c

$$u = \frac{c}{\rho_0} \tag{3.5}$$

where ρ_0 density in absolute dry condition

the relation of the moisture content in the cross section can be determined by

$$\frac{\partial u}{\partial t} = \nabla \cdot (\mathbf{D} \cdot \nabla u) \tag{3.6}$$

As input value for the determination of the moisture content, the diffusion coefficient $\mathbf{D}$ is often given as a part of the rheological model.

Besides the moisture diffusion within the cross section, the boundary conditions must be defined:

$$\frac{q_n}{\rho_0} = S \cdot (u_{\text{air}} - u_{\text{surface}}) \tag{3.7}$$

where S surface emissivity

and

$$D \cdot \frac{\partial u}{\partial n} = S \cdot (u_{\text{air}} - u_{\text{surface}}) \tag{3.8}$$

where n y,z
direction of the normal vector of the surface

Comparable to the coefficient $\mathbf{D}$ the surface emissivity $\mathbf{S}$ is often given for the rheological model, since these models are calibrated to the moisture content and to the evaluation of the moisture content based on the diffusion coefficient $\mathbf{D}$ and the surface emissivity $\mathbf{S}$, respectively.

3.2 Rheological models

3.2.1 General

In the last few years a lot of models have been developed. So the presented models are just an extract thereof. Since the initial starting point of this study was the comparison of the modeling of the long-term behavior of timber-concrete composite slabs and solving the question, why there is such a variability in the evaluated values of these timber-concrete composite models (see Grosse et al. [18], Fragiacomo [15], Bou Saïd [3] and [48]), mainly those models are compared, which were used for the description of the long-term behavior of these composite slabs.

3.2.2 Model according to Toratti [51]

General equations Toratti [51] proposed several ways of modeling the time-dependent behavior (model A, B, C, D and refined). Since the original aim of the studies was the discussion of the long-term behavior of timber-concrete composite structures and Fragiacomo [15] used model B without the extensions of model D and refined model D, the evaluation is mainly done by model B.

Model B according to Toratti [51] consists of six Kelvin-Voigt-bodies for the pure creep, one Kelvin-Voigt-body for the mechano-sorptive creep and finally an element, considering the hydroexpansion (see Fig. 3.2). The time-dependent strain therefore consists of five parts:

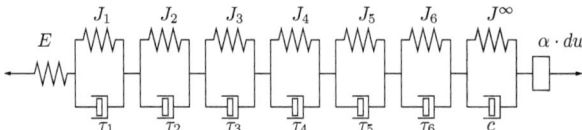

Figure 3.2: Model according to Toratti [51]

- elastic strain due to change of the Modulus of Elasticity and due to change of the stresses

$$\varepsilon_{\text{elastic}}(t) = \int \left(J_0 \cdot \dot{\sigma} + \dot{J}_0 \cdot \sigma \right) \, dt \tag{3.9}$$

where σ stress
J_0 elastic compliance
       $= 1/E(u)$
$E(u)$ Modulus of Elasticity

- normal creep strain

$$\varepsilon_{\text{creep}}(t) = J_0(u_{ref}) \cdot \int \sum_{i=1}^{6} J_n \cdot \left(1 - e^{-\frac{t-t'}{\tau_n}} \right) \cdot d\sigma(t') \tag{3.10}$$

3.2 Rheological models

where $J_0(u_{ref})$ elastic compliance for the reference moisture content
J_n, τ_n parameters (comp. Tab. 3.1)
σ stress

Table 3.1: Parameters for the model according to Toratti [51]

	J_n	τ_n [days]
1	0.0686	0.01
2	-0.0056	0.1
3	0.0716	1
4	0.0404	10
5	0.2073	100
6	0.5503	5000

The parameters of Tab. 3.1 are determined by fitting the model to the following power-function proposed by Toratti [51].

$$J(t,0) = J_0(u) + J_0(u_{ref}) \cdot \left(\frac{t}{t_d}\right)^k \tag{3.11}$$

where t current time
t_d 29500 days
k = 0.21
u_{ref} = 0.2

- mechano-sorptive creep strain: The stress-strain relation of this Kelvin-Voigt-body for the mechano-sorptive creep is given by

$$\varepsilon_{ms}(t) = J^\infty \cdot \int_0^t \left(1 - e^{-c \cdot \int_t^{t'} |du(t'')|}\right) d\sigma(t') \tag{3.12}$$

where ε_{ms} mechano-sorptive creep strain
u moisture content
σ stress
J^∞ = $0.7 \cdot J_0(u_{ref})$
c = 2.5

- the inelastic strain due to moisture variation: The inelastic strain due to moisture variation can be determined by the following equation

$$\varepsilon_u(t) = \int_0^t (a - b \cdot \varepsilon(t')) \, du(t') \tag{3.13}$$

where ε_u shrinkage/swelling
u moisture content
b = 1.3
a = 0.00625

Within this equation the influence of the total strain on the inelastic strain due to moisture variation is considered by the term $-b \cdot \varepsilon(t')$.

- influence of the stress on the mechano-sorptive creep (extension of model B for model D): In order to improve the model, the combined model is introduced (comp. Toratti

[51]), where the irrecoverable strain is considered by

$$\varepsilon_{irrec} = J_0(u_{ref}) \cdot e \cdot \int_0^t \sigma(t') \, |du(t')| \qquad \text{for compression only} \tag{3.14}$$

where $e = 0.1$

In order to improve the model for low humidities, Toratti [51] finally proposed to use an effective moisture rate only for compression stresses (extension of model B and model D for the refined model).

$$\dot{u}_{\text{eff}} = \dot{u} \cdot e^{200 \cdot (u - 0.18)} \tag{3.15}$$

For the numerical evaluation within the scope of this study, this equation has been modified to the following equation for reasons of numerical stability

$$\dot{u}_{\text{eff}} = \begin{cases} \dot{u} \cdot e^{200 \cdot (u - 0.18)} & \text{for } u \leq 18\% \\ \dot{u} & \text{for } u > 18\% \end{cases} \tag{3.16}$$

Because the elastic compliance J_0 depends on the Modulus of Elasticity, Toratti [51] proposes the following relation between the Modulus of Elasticity and the moisture content

$$E(u) = E(u = 0) \cdot (1 - 1.06 \cdot u) \tag{3.17}$$

Since the Modulus of Elasticity is often not given for the moisture content of $u = 0$, Eq. (3.17) is transformed to

$$E(u) = E_0 \cdot \frac{1 - 1.06 \cdot u}{1 - 1.06 \cdot u_0} \tag{3.18}$$

Finally, Toratti [51] proposed an extension for modeling the material non-linearity, the tertiary creep phase and the final rupture by introducing the concept of damage accumulation. This extension is not included in the following studies, because these studies aim at the creep on the load level of the serviceability limit state.

Coefficients for the moisture transportation For the evaluation of the moisture content in dependence on the surrounding conditions the diffusion coefficient D is given by

$$D = 1.2 \cdot 10^{-10} \cdot e^{2.28 \cdot u} \quad [\text{m}^2/\text{s}] \tag{3.19}$$

where u moisture content

and the surface emissivity S can be described by

$$S = 1.3 \cdot 10^{-7} \quad [\text{m/s}] \tag{3.20}$$

The equilibrium moisture content can be evaluated by

$$u_{RH} = \frac{0.01 \cdot RH}{-0.00084823 \cdot RH^2 + 0.11665 \cdot RH + 0.38522} \tag{3.21}$$

where RH relative humidity

3.2.3 Model according to Hanhijärvi [22]

General equations   The rheological model according to Hanhijärvi [22] consists of nine Maxwell-bodies and one spring in parallel order. A component for the hygro-expansion is added to each parallel set of elements (see Fig. 3.3). The change of the strain depending

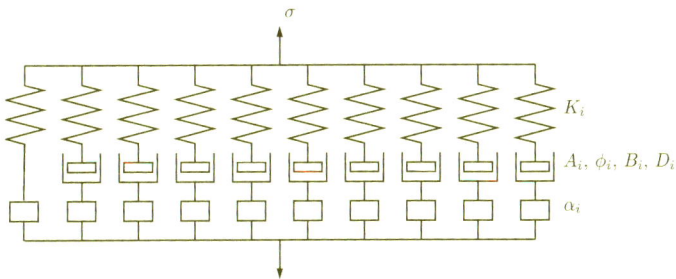

Figure 3.3: Model according to Hanhijärvi [22]

on time is given by

$$\dot{\varepsilon} = A \cdot sinh\left(\phi \cdot \sigma + B \cdot tanh\left(D \cdot \dot{h}\right)\right) + \alpha_h \cdot \dot{h} \tag{3.22}$$

where A, B, D, ϕ parameters (comp. Tab. 3.2)
α_h coefficient of hygroexpansion
σ stress in the element
h relative vapor pressure

Table 3.2: Parameters for the rheological model according to Hanhijärvi [22]

K_0	$= 0.25 \cdot E$	$K_1 \ldots K_9$	$= 0.0833 \cdot E$		
A_0	$= 0$	A'_1	$= \frac{0.002}{E}$ [MPa/h]	A'_i	$= \frac{A'_{i-1}}{3}$
ϕ_0	$= 0$	ϕ_1	$= 12$ [1/MPa]	ϕ_i	$= \frac{\phi_{i-1}}{1.3}$
B_0	$= 0$	$B_1 \ldots B_9$	$= 8.5$		
D_0	$= 0$	$D_1 \ldots D_9$	$= 100$ [h]		
α'_h	$= 0.0016$				
b_A	$= 400$	b_α	$= 180$	E	$=$ Modulus of Elasticity

In order to consider the different behavior for tension and compression, the coefficients A and α_h are modified in dependence on the mechanical strain ε_m. The mechanical strain is the total strain without hydroexpansion.

$$\begin{array}{ll} \varepsilon_m \leq 0 & \left\{ \begin{array}{ll} A &= A' \cdot (1 - b_A \cdot \varepsilon_m) \\ \alpha_h &= \alpha'_h \cdot (1 - b_\alpha \cdot \varepsilon_m) \end{array} \right. \\ \varepsilon_m > 0 & \left\{ \begin{array}{ll} A &= A' \cdot e^{-b_A \cdot \varepsilon_m} \\ \alpha_h &= \alpha'_h \cdot e^{-b_\alpha \cdot \varepsilon_m} \end{array} \right. \end{array} \tag{3.23}$$

The dependence of the Modulus of Elasticity on the moisture content is given by

$$E(u) = E_0 \cdot (1 - 1.06 \cdot (u - u_0)) \tag{3.24}$$

where E Modulus of Elasticity
u moisture content
subscript 0 initial state

Coefficients for the moisture transportation The diffusion coefficient D is proposed to be

$$D = 8.0 \cdot 10^{-11} \cdot e^{4.0 \cdot u} \ \text{m}^2/\text{s} \tag{3.25}$$

where u moisture content

The surface emissivity S is proposed to be

$$S = 3.2 \cdot 10^{-8} \cdot e^{4.0 \cdot u} \ \text{m/s} \tag{3.26}$$

where u moisture content

Finally the equilibrium moisture content in this model is proposed to be

$$u_{air} = 0.01 \cdot \left(\frac{-T \cdot ln(1-h)}{0.13 \cdot \left(1 - \frac{T}{647.1}\right)^{-6.46}} \right)^{\frac{1}{110 \cdot T^{-0.75}}} \tag{3.27}$$

where T temperature in K
h relative vapor pressure
 $= 0.01 \cdot RH$

For the determination of the relative vapor pressure in Eq. (3.22), the inverse function of Eq. (3.27) is used, where the variable u_{air} is replaced by the current moisture content at each location of the cross section.

3.2.4 Model according to Becker [2]

General equations The rheological model according to Becker [2] consists of four Kelvin-Voigt-bodies, which represent the normal creep and additionally one Kelvin-Voigt-body, which represents the mechano-sorptive creep. In order to consider the inelastic strains due to moisture variation and due to non-linear creep, additional elements are added (see Fig. 3.4). Therefore the time-dependent strain is split in four parts:

3.2 Rheological models

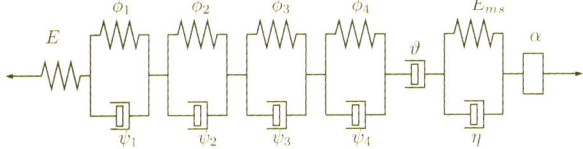

Figure 3.4: Rheological model according to Becker [2]

- elastic strains due to the change of material properties, caused by a moisture variation

$$\varepsilon_{\text{elastic}}(t) = \int \frac{\dot{\sigma}(t)}{E(u(t))} + \frac{\sigma(t)}{\dot{E}(u(t))} \, dt \qquad (3.28)$$

where σ stress
$E(u)$ Modulus of Elasticity

- normal creep strain:

$$\varepsilon_{\text{creep}}(t) = \int \sum_{i=1}^{4} \phi_i \cdot \left(1 - e^{-\frac{t-t'}{\psi_i}}\right) \cdot d\sigma(t') \qquad (3.29)$$

ϕ_i, ψ_i parameters (comp. Tab. 3.3)
σ stress

Table 3.3: Parameters for the model according to Becker [2]

	ϕ_i	ψ_i [h]
1	0.08	15
2	0.08	400
3	0.22	4000
4	0.22	28000

- mechano-sorptive creep strain: For the mechano-sorptive creep strain, a Kelvin-Voigt-body is used. Therefore the differential equation can be set up as

$$\frac{d\varepsilon_{ms}}{du} = \frac{\sigma(t)}{\eta_{ms}} - \frac{E_{ms} \cdot \varepsilon_{ms}}{\eta} \qquad (3.30)$$

where E_{ms} stiffness of the spring
η damping coefficient
$\sigma(t)$ stress
ε_{ms} strain due to mechano-sorptive creep

The damping coefficient is given by

$$\eta = \frac{E}{\alpha_L} \cdot 1.25 \cdot 10^{-3} \qquad (3.31)$$

where E Modulus of Elasticity
        α_L swelling/shrinkage coefficient

In contrast to the pure creep formulation, the parameter of the spring depends on the moisture history.

$$E_{ms} = \frac{\eta}{\Delta u} \qquad (3.32)$$

where Δu $= u_{\max} - u_{\min}$
        $u_{\max}$ maximum moisture content
        $u_{\min}$ minimal moisture content

For the implementation it is assumed that Δu is the maximum difference within the interval $t = 0$ and $t = t$. Besides that, Eq. (3.30) is modified so, that only absolute changes of the moisture influence the equation.

$$\frac{d\varepsilon_{ms}}{|du|} = \frac{\sigma(t)}{\eta_{ms}} - \frac{E_{ms} \cdot \varepsilon_{ms}}{\eta} \qquad (3.33)$$

This is essential because following Eq. (3.30) a negative mechano-sorptive creep strain would arise in an element subjected to tension in the drying process. This means a creep strain opposite to the stress direction would arise, resulting in a reduced deformation caused by mechano-sorptive creep.

- non-linear strain by reaching the limit of proportionality: In order to consider the limit of proportionality an additional damper is introduced. Therefore the strains, caused by this element, can be set up

$$\varepsilon_{nl} = \frac{1}{E} \cdot \int_0^t \vartheta_{nl} \, dt \qquad (3.34)$$

The parameter ϑ_{nl} can be determined to be

$$\vartheta_{nl} = \begin{cases} \pm \alpha_{nl} \cdot (\sigma - \sigma_{LoP})^{\beta_{nl}} & \text{for } \sigma > \sigma_{LoP} \\ 0 & \text{for } \sigma \leq \sigma_{LoP} \end{cases} \qquad (3.35)$$

where α_{nl} $= 0.0014$
        β_{nl} $= 2$
        σ stress
        σ_{LoP} limit of proportionality

The sign of the coefficient ϑ_{nl} has to be adjusted so, that the sign of the coefficients fits the direction of the stress.

The limits of proportionality can be determined to be

$$\frac{\sigma_{LoP}(u)}{f(u = 12\%)} = A \cdot e^{-B \cdot u^C} \qquad (3.36)$$

where $f(u = 12\%)$ strength for a moisture content of 12%
        A, B, C coefficients according to Tab. 3.4

- inelastic strain due to moisture variation: The part of shrinkage and swelling can be

3.2 Rheological models

Table 3.4: Coefficients A, B and C for the determination of the limit of proportionality

	compression	tension
A	0.1437	0.1305
B	-0.04111	-0.4119
C	-1.5162	-0.6416

determined by

$$\varepsilon_u = \int \bar{\alpha}_L \, du \qquad (3.37)$$

where $\bar{\alpha}_L$ shrinkage/swelling coefficient

In order to consider the different inelastic strains due to moisture content in dependence on the strain, the shrinkage/swelling coefficient is modified in dependence on the elastic and viscoelastic strain (= pure creep strain) and the strain due to the non-linearity.

$$\bar{\alpha}_L = \begin{cases} \alpha_L \cdot (1 - b_\alpha \cdot \varepsilon_{\text{mech}}) & \text{for } \varepsilon_{\text{mech}} \leq 0 \\ \alpha_L \cdot e^{-b_\alpha \cdot \varepsilon_{\text{mech}}} & \text{for } \varepsilon_{\text{mech}} > 0 \end{cases} \qquad (3.38)$$

where α = 0.008 ± 0.002
in the following the variability of the shrinkage/swelling coefficient is neglected. So $\alpha = 0.008$ is used.
b_α = 180
$\varepsilon_{\text{mech}}$ = $\varepsilon_{\text{elastic}} + \varepsilon_{\text{pure creep}} + \varepsilon_{\text{non-linearity}}$

So neither the hydroexpansion strain nor the strain caused by mechano-sorptive creep influence the coefficient $\bar{\alpha}_L$.

For the dependence of the Modulus of Elasticity on the moisture content the following proposal is given

$$E(u) = E(u = 12\%) \cdot (1 - 1.5 \cdot (u - 0.12)) \qquad (3.39)$$

To generalize the input values, Eq. (3.39) was transformed to

$$E(u) = E_0 \cdot \frac{1 - 1.5 \cdot (u - 0.12)}{1 - 1.5 \cdot (u_0 - 0.12)} \qquad (3.40)$$

where $E(u)$ Modulus of Elasticity in dependence on the moisture content
E_0 initial Modulus of Elasticity
u current moisture content
u_0 initial moisture content

Coefficients for the moisture transportation In order to determine the moisture content the diffusion coefficient is proposed to be

$$D = 0.5 \cdot \left(1 - \frac{\rho_0 - 420}{420} \cdot 2\right) \cdot e^{4.0 \cdot u} \qquad [\text{mm}^2/\text{h}] \qquad (3.41)$$

For the determination of the surface moisture content following equation is given:

$$u_{Surface} = u_{Surface} + (u_{RH} - u_{Surface}) \cdot \left(1 - e^{-\beta \cdot t}\right) \qquad (3.42)$$

where u_{RH} equilibrium moisture content according to
β coefficient according to Eq. (3.43)

The coefficient β is evaluated by

$$\beta = \frac{0.03}{1 - e^{-\frac{d^2}{40}}} \qquad (3.43)$$

where d thickness in mm

For the numerical implementation the coefficient β is rounded to 0.03, because the thickness of the common cross-section dimensions is much larger than 40mm (see Fig. 3.5).

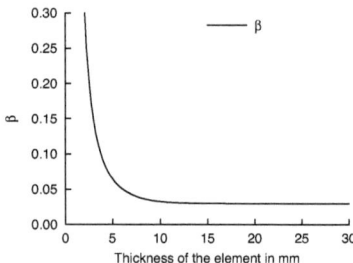

Figure 3.5: Influence of the thickness on the coefficient β for the evaluation of the equilibrium moisture content

Besides that Eq. (3.42) is transformed to a surface emissivity value in order to fit the equations in Chap. 3.1.

$$S = \left(1 - e^{-0.03 \cdot dt}\right) \cdot \frac{dx}{dt} \qquad (3.44)$$

The equilibrium moisture content is determined to be

$$u(RH) = 0.113 \cdot RH^{0.54} + 0.192 \cdot e^{-0.5 \cdot (2.7 \cdot (RH-1)-1)^2} + 0.09 \cdot e^{-0.5 \cdot (20.5 \cdot (RH-1)-1)^2} \qquad (3.45)$$

where RH relative humidity as an absolute value

3.2.5 Model according to Mårtensson [38]

General equations In the model according to Mårtensson [38] the time-dependent strain depends on five different parts (see Fig. 3.6):

3.2 Rheological models

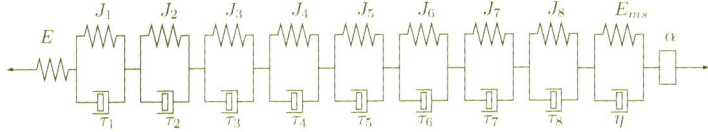

Figure 3.6: Model according to Mårtensson [38]

- elastic strain rate:

$$\dot{\varepsilon}_{\text{elastic}} = \frac{\dot{\sigma}}{E} \tag{3.46}$$

- normal creep strain: To evaluate the pure creep strain, eight Kelvin-Voigt-bodies are used.

$$\varepsilon_{\text{creep}}(t) = \int \sum_{i=1}^{8} J_n \cdot \left(1 - e^{-\frac{t-t'}{\tau_n}}\right) \cdot d\sigma(t') \tag{3.47}$$

For the parameters τ_n and J_n two different sets are given. However, the set in Tab. 3.5 (creep data 2) is used for the simulations. In order to adjust this normal creep to different moisture contents, a material time ξ is introduced. This material time can be determined by the following equation

$$d\xi = \frac{dt}{a(u)} \tag{3.48}$$

where $a(u)$ moisture shift factor according to Fig. 3.7

The moisture shift factor $a(u)$ is given in Fig. 3.7. Since it was not possible to set up an equation, showing this moisture shift factor in a sufficiently exact way, the moisture shift factor is linearized between the intervals given in Tab. 3.6

- mechano-sorptive creep strain: For the mechano-sorptive creep strain the following equation is proposed

$$\dot{\varepsilon}_{ms} = m \cdot \sigma \cdot |\dot{u}| \tag{3.49}$$

Table 3.5: Material parameters τ_n and J_n (creep data 2, see Mårtensson [38])

	τ_i in h	J_n in MPa^{-1}
1	0.01	$1.0140 \cdot 10^{-6}$
2	0.1	$3.6140 \cdot 10^{-6}$
3	1	$2.0960 \cdot 10^{-6}$
4	10	$5.1900 \cdot 10^{-7}$
5	100	$1.6160 \cdot 10^{-6}$
6	1000	$2.3180 \cdot 10^{-5}$
7	10000	$2.9150 \cdot 10^{-6}$
8	100000	$3.6290 \cdot 10^{-5}$

Table 3.6: Intervals for the moisture shift factor a

u	a
0	68.5277
0.0188	63.0075
0.0498	48.9745
0.0751	32.181
0.0879	22.5208
0.1008	14.7983
0.1253	8.0496
0.1601	3.4034
0.209	1.1185
0.2587	0.3451
0.2953	0.1621
0.3159	0.1021

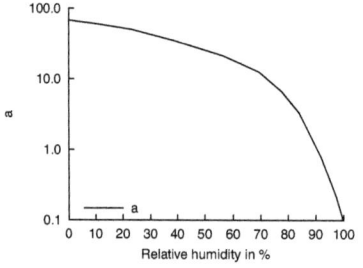 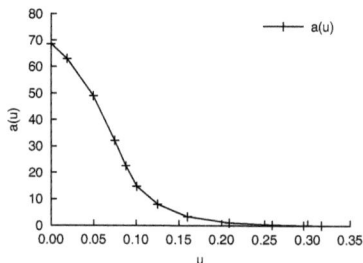

(a) Dependence on relative humidity (from Mårtensson [38])

(b) Dependence on the moisture content

Figure 3.7: Moisture shift factor a

where $\dot{u}$ temporal change of the moisture content
σ stress
m coefficient according to Eq. (3.50)

The coefficient for the mechano-sorptive creep is dependent on the maximum moisture variation within the mechano-sorptive period. It can be determined to be

$$m = \begin{cases} m_0 & \text{for} \quad u < u_{min} \text{ or } u > u_{max} \\ m_0 \cdot e^{\frac{\varepsilon_{mst}}{\varepsilon_{mst}-\varepsilon_{ms\infty}}} & \text{for} \quad u_{min} \leq u \leq u_{max} \end{cases} \quad (3.50)$$

3.2 Rheological models

where ε_{mst} total mechano-sorptive strain (comp. Eq. (3.53)) including the mechano-sorptive recovery (comp. Eq. (3.51))
$\varepsilon_{ms\infty}$ mechano-sorptive creep limit according to Tab. 3.7
u_{min} minimal moisture content within the mechano-sorptive period (comp. Eq. (3.54))
u_{max} maximum moisture content within the mechano-sorptive period (comp. Eq. (3.54))

Table 3.7: Material properties for mechano-sorptive creep and recovery (see Mårtensson [38])

	m_0 (MPa)$^{-1}$	$\varepsilon_{ms\infty}$	λ	k
Leivo, spruce in bending	$2.0 \cdot 10^{-4}$	$3.0 \cdot 10^{-4}$	0.1	1.7
Taylor, glulam, European whitewood	$2.0 \cdot 10^{-4}$	$5.5 \cdot 10^{-4}$	0.1	1.7
Mårtensson [38]	$2.0 \cdot 10^{-4}$	$5.5 \cdot 10^{-4}$	0.1	1.7

- mechano-sorptive recovery:

$$\dot{\varepsilon}_{msr} = -L \cdot \left| \frac{\sigma^* - \sigma}{\sigma^*} \right| \cdot \varepsilon_{mst} \cdot \dot{u} \qquad (3.51)$$

where ε_{mst} total mechano-sorptive strain (comp. Eq. (3.53)) including the mechano-sorptive recovery
σ stress
σ^* average stress within the mechano-sorptive period (comp. Eq. (3.55))
$\dot{u}$ change of moisture content
L coefficient according to Eq. (3.52)

The coefficient L is given by

$$L = \begin{cases} \frac{\lambda}{u_f} & \text{for} \quad \dot{u} > 0 \quad \text{and} \quad \begin{cases} \sigma(t) < \sigma^*(t) & \text{when} \quad \sigma^*(t) > 0 \\ \sigma(t) > \sigma^*(t) & \text{when} \quad \sigma^*(t) < 0 \end{cases} \\ 0 & \text{in all other cases} \end{cases} \qquad (3.52)$$

where λ coefficient according to Tab. 3.7
u_f fiber saturation point, assumed at 0.32

With Eq. (3.49) and Eq. (3.51) the total strain due to mechano-sorptive can be determined by

$$\varepsilon_{mst} = \int_0^t \dot{\varepsilon}_{ms} \, dt + \int_0^t \dot{\varepsilon}_{msr} \, dt \qquad (3.53)$$

The mechano-sorptive period is defined as the time span, when

$$\varepsilon_{mst}(t_{ms}) = 0 \qquad (3.54)$$

where ε_{mst} total mechano-sorptive strain (comp. Eq. (3.53))

That means, that the mechano-sorptive period has finished, when the total mechano-sorptive creep equals to 0.

The average stress within the mechano-sorptive period is defined as

$$\sigma^*(t) = \frac{1}{t - t_{ms}} \cdot \int_{t_{ms}}^{t} \sigma \, dt \qquad (3.55)$$

It is assumed, that the material has no memory of the previous mechano-sorptive period, so the maximum and the minimum moisture content are set to the current moisture content.

$$u_{min}(t_{ms}) = u \qquad (3.56)$$
$$u_{max}(t_{ms}) = u \qquad (3.57)$$

- inelastic strain due to shrinkage/swelling: The inelastic strain due to shrinkage and swelling can be described by a modified shrinkage/swelling coefficient, which considers the dependence of the shrinkage/swelling rate on the strain.

$$\dot{\varepsilon} = (\alpha - \Delta\alpha) \cdot \dot{u} \qquad (3.58)$$

where α shrinkage/swelling coefficient
       $\Delta\alpha$ modification due to external strain

For the shrinkage/swelling coefficient α several values are proposed (comp. Tab. 3.8). However, for this study, a value of 0.0054 as proposed by Mårtensson [38] is chosen.

Table 3.8: Proposed values for the shringake/swelling coefficient

	Mårtensson [38] spruce	Mohager [39] pine	Hunt [29] scots pine	Hunt [29] ponterosa pine	Leivo [35] spruce	Taylor et al. [50] glulam European white-wood
α_L	$5.4 \cdot 10^{-3}$	$6.1 \cdot 10^{-3}$	$4.6 \cdot 10^{-3}$	$5.6 \cdot 10^{-3}$	$4.0 \cdot 10^{-3}$	$6.1 \cdot 10^{-3}$

In order to consider the influence of the strain on the swelling/shrinkage coefficient, the coefficient $\Delta\alpha$ can be determined in the following way

$$\Delta\alpha = k \cdot p \cdot \varepsilon_{\text{elastic}} + k \cdot (\varepsilon_{\text{normal creep}} + \varepsilon_{mst}) \qquad (3.59)$$

where $\varepsilon_{\text{elastic}}$ elastic strain
       $\varepsilon_{\text{normal creep}}$ normal creep according to Eq. (3.47)
       ε_{mst} total mechano-sorptive creep according to Eq. (3.53)
       k,p coefficients

For the coefficient p it is proposed to use a value of 0.25 and for the coefficient k a value of 1.7 respectively (see Mårtensson [38]).

Concerning the dependence of the Modulus of Elasticity on the moisture content, no formula but different values are given (see Tab. 3.9). A relation between the moisture content and

3.2 Rheological models

Table 3.9: Modulus of Elasticity for three relative humidities

Modulus of Elasticity MPa	Mårtensson [38] spruce	Mohager [39] pine in tension	Mohager [39] pine in bending	Hunt [29] scots pine	Hunt [29] ponterosa pine	Leivo [35] spruce	Taylor et al. [50] glulam European white-wood
30% RH	13200	10600	12600	9700	11800	13500	12600
60% RH	11600	10000	11000	9000	10800	12600	11000
90% RH	10800	8600	10200	7500	9200	11000	10200

the Modulus of Elasticity can be derived from the average values of Tab. 3.9 and a linear fit,

$$E(u) = E_0 \cdot \frac{1 - 1.58 \cdot u}{1 - 1.58 \cdot u_0} \tag{3.60}$$

Table 3.10: Parameters, used in the simulations

creep data	2
m_0 in $(\% \text{ MPa})^{-1}$	$2.0 \cdot 10^{-6}$
$\varepsilon_{ms\infty}$	$5.5 \cdot 10^{-4}$
λ	0.1
k in $\%^{-1}$	0.017
$a(u)$	according to Fig. 3.7
$E(u)$	according to Eq. (3.60)
$\alpha_{\parallel}$	$5.4 \cdot 10^{-5}$

For the simulations the values given in Tab. 3.10 are used.

Coefficients for the moisture transportation Within the scope of Mårtensson [38], no explicit values for the diffusion coefficient and the surface emissivity are given, since the moisture content is evaluated by JAM developed by Arfvidsson [1].

For the equilibrium moisture content, Fig. 3.8 is given. So the following equation is set up accordingly to the values:

$$\begin{aligned} u(RH) = & 0.2439353898 \cdot \left(\frac{RH}{100\%}\right)^6 + 0.1276829904 \cdot \left(\frac{RH}{100\%}\right)^5 \\ & + 0.09340840267 \cdot \left(\frac{RH}{100\%}\right)^4 - 0.464828494 \cdot \left(\frac{RH}{100\%}\right)^3 + \\ & 0.1208946628 \cdot \left(\frac{RH}{100\%}\right)^2 + 0.1948280174 \cdot \left(\frac{RH}{100\%}\right) \end{aligned} \tag{3.61}$$

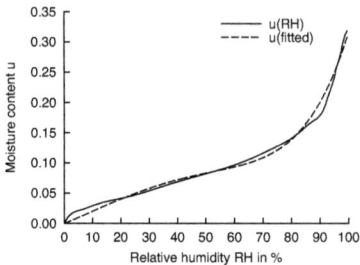

Figure 3.8: Relation between relative humidity RH and moisture content u

For the numerical implementation it is assumed that the diffusion coefficient and the surface emissivity used in the model according to Toratti [51] are valid.

3.3 Summary

The presented models differ in their build-up as well as in their parameters. Therefore different responses are expected. Do the models differ significantly? For this reason all the presented models are implemented for the use in *kriHo*. In order to ensure a proper implementation, some results given for the particular model are re-evaluated with *kriHo* (see Appendix A).

In order to answer the question, whether the models differ significantly, theoretical as well as numerical studies are performed in the next chapter.

4 Comparison of the creep coefficients

4.1 General comparison of the models

All introduced models describe the time-dependent behavior with respect to the variation of the moisture content. Since the mathematical formulations differ, different results are expected. Whereas the models according to Toratti [51], Becker [2] and Mårtensson [38] are serial chains of Kelvin-Voigt-bodies (see Fig. 4.1), the model according to Hanhijärvi

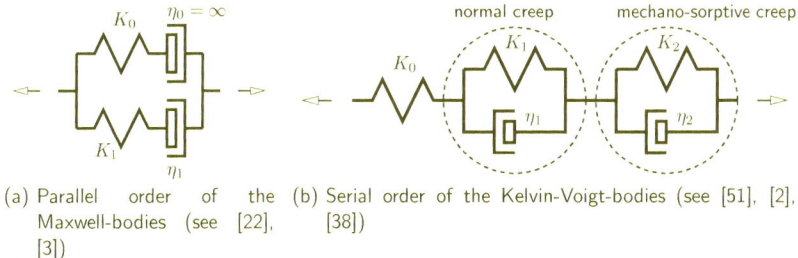

(a) Parallel order of the Maxwell-bodies (see [22], [3])

(b) Serial order of the Kelvin-Voigt-bodies (see [51], [2], [38])

Figure 4.1: Differences in the build-up of the models

[22] consists of parallel Maxwell-bodies. Additionally, in the model of Hanhijärvi [22] one Maxwell-element only consists of a spring and an element for the swelling/shrinkage. So the total creep is limited to a theoretical creep limit of $\varphi = 4$ (see Eq. (3.22)). In this model it is irrelevant, whether the creep is caused by mechano-sorptive creep or by normal creep (see Fig. 4.2). The stresses in this model are shifted from the weaker to the stronger elements. Therefore it is expected, that in the model according to Hanhijärvi [22] the difference between pure creep deformation and the combination of pure creep and mechano-sorptive creep can be neglected, provided the considered period is long enough. If the period of 50 years is long enough, the determination of the creep coefficients would be simplified, because no influences of the surrounding conditions are expected and the creep coefficient would depend only on the stress level.

Since the other models consist of serial chains of Kevin-Voigt-bodies, the creep limit is caused by the spring within the single Kelvin-Voigt-body. These creep limits of the single Kelvin-Voigt-bodies are summed up to give the global creep limit. Besides that, the different types of creep are added. Since the total creep depends on the different springs of the Kelvin bodies, no general creep limit can be given. In a situation, where the mechano-sorptive creep is decisive for the time-dependent behavior, the theoretical creep limit is more or less the creep limit of the spring in the Kelvin-body, respecting the mechano-sorptive creep. On the other hand, in a situation, where the normal creep strain determines the time-dependent

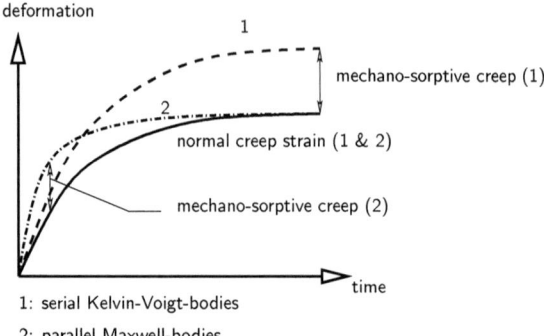

Figure 4.2: Principle course of the creep strain of both typical models for normal and mechano-sorptive creep

strain, the creep limit is more or less the creep limit of the Kelvin-bodies respecting the normal creep.

This aforementioned influence of the type of the model contradicts the common use of the models, since the parameters of the single models can be transferred to each other. Therefore only in normal creep there is no difference, whether serial Kelvin-Voigt-bodies or parallel Maxwell-bodies are used ("Duality of the differential equations"). Since the two independent parameters time t and moisture variation Δu determine the time-dependent behavior in these models, the duality of the differential equations is not valid any more. Therefore the parameters of serial Kelvin-Voigt-bodies can not be transformed into the parameters of parallel Maxwell-bodies (for details see Appendix D).

Besides the different modeling of the time-dependent behavior, the moisture content determination differs for the single models. Since the models are developed with respect to a certain moisture determination, the parameters for the evaluation of the moisture content proposed in the models are used, although it results in different average moisture contents, different moisture variations and different equilibrium moisture contents.

4.2 Constant surrounding conditions

As a first step, the different creep coefficients according to the rheological models are determined. The following boundary conditions are assumed, which are comparable to creep tests by Hoyle et al. [28]:

- single span girder with a span of 80cm and a cross section of 8.9cm × 8.9cm
- constant moisture content of 10%
- Modulus of Elasticity equal 14000N/mm^2 at a moisture content of 0%
- edge stresses equal 13.2N/mm^2

4.2 Constant surrounding conditions

- creep coefficient φ

$$\varphi = \frac{w(t = 50 \text{ years})}{w_{\text{elastic}}} - 1 \tag{4.1}$$

where w mid-span deflection of the beam

With these boundary conditions, the deflections and the creep coefficients according to the models are determined. As seen in Tab. 4.1 and Fig. 4.3(a) the models of Hanhijärvi [22] and Toratti [51] evaluate creep coefficients after 50 years within the range of 1.04 and 1.12, whereas the models according to Becker [2] and to Mårtensson [38] result in creep coefficients in the range of 0.52 and 0.6.

Table 4.1: Creep coefficient for constant loading, constant moisture content of 10% and an edge stress of 13.2N/mm^2

Time	hours	DoL	Eurocode 5 [14]	Becker [2]	Hanhijärvi [22]	Mårtensson [38]	Toratti [51]
6 month	4320	middle	0.25	0.31	0.41	0.17	0.39
10 years	87600	long	0.5	0.59	0.80	0.42	0.75
50 years	438000	permanent	0.6	0.60	1.13	0.52	1.04

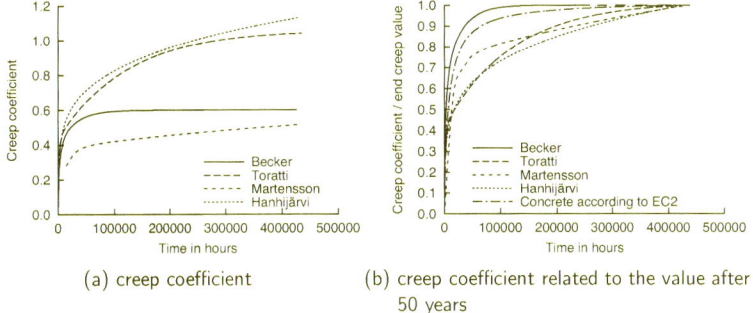

(a) creep coefficient

(b) creep coefficient related to the value after 50 years

Figure 4.3: Creep coefficient for constant loading, constant moisture content of 10% and an edge stress of 13.2N/mm^2

Besides the difference in the absolute value, the temporal development is not equal in all rheological models. The model according to Becker [2] reaches its end value within the first ten years, whereas the creep coefficients of the other models continue to increase (see Fig. 4.3(b)). These other three models show a more or less similar development of the creep strain.

In Fig. 4.4 the creep coefficients evaluated by the models after 50 years of constant loading in constant surrounding conditions are shown. The stress level influences only the creep

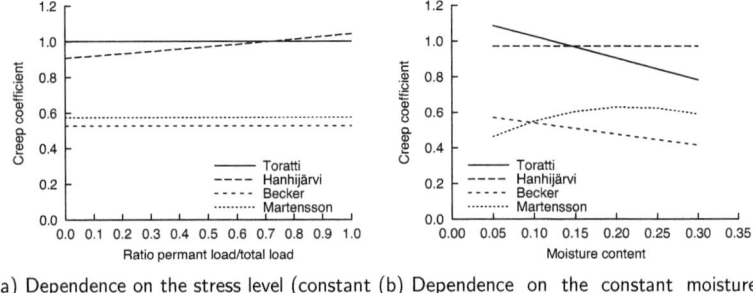

(a) Dependence on the stress level (constant moisture content $u_0 = 12\%$)

(b) Dependence on the constant moisture content (stress level 50%)

Figure 4.4: Comparison of the creep coefficients after 50 years

coefficient according to Hanhijärvi [22]. However, this influence seems to be negligible since it varies between 0.9 and 1.1. In contrast to the stress level, the initial moisture content obviously influences the effective creep coefficient in the models according to Toratti [51], Becker [2] and Mårtensson [38]. In the model according to Hanhijärvi [22] this influence can be neglected.

To sum up the results of the evaluation of the creep coefficient for a period of 50 years in constant climate the following conclusions can be drawn:

- The creep coefficients after 50 years depend on the model. In constant climate the values are between 0.5 and 1.1 for the same cases (see Fig. 4.3 and Fig. 4.4). So no unique value of the creep deformation evaluated by the models can be given.

- Only the model according to Hanhijärvi [22] leads to a dependence of the creep deformation on the stress level. The constant moisture content does not influence the creep coefficient in this model, whereas in the other models, the effective creep coefficient is a function of the moisture content but not of the stress level (see Fig. 4.4).

- In the model according to Toratti [51] and Becker [2] the creep coefficient decreases with increasing moisture content (see Fig. 4.4(b)). The elastic deformation increases with increasing moisture content because the Modulus of Elasticity decreases. The creep strain depends on the Modulus of Elasticity at a moisture content of 20% (see $J_{u,ref}$ in Eq. (3.10)) and therefore is independent on the moisture content. As the creep coefficient is the ratio between creep strain and elastic strain, the creep coefficient finally decreases.

In the model according to Hanhijärvi [22] the basic creep coefficient is independent of the moisture content, whereas in the model according to Mårtensson [38] the creep coefficient increases with increasing moisture content.

4.3 Effective creep coefficients in variable climate

For comparing the rheological models in variable conditions the following assumptions are made:
- the average relative humidity is 65%,
- the difference between maximum and minimum relative humidity is 30%, which is equivalent to an amplitude of $\Delta RH = 15\%$,
- sinusoidal course of the relative humidity
- a constant temperature of $18°C$ over the whole period of time
- the thickness of the cross section is 10cm and the height is 20cm and
- the ratio between permanent load and total load is 100%,

if the values are not explicitly defined.

Fig. 4.5 shows the influence of the amplitude of the annual course of the relative humidity. As can be seen, the creep coefficient according to the model of Hanhijärvi [22] is hardly

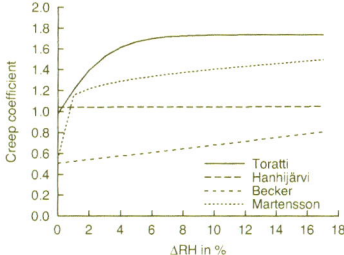

Figure 4.5: Influence of the amplitude of the relative humidity on the creep coefficient

influenced, because in this model the total creep strain is limited due to the parallel order of the Maxwell-bodies. So the creep coefficient does not depend on the case whether the creep strain is caused by mechano-sorptive creep or normal creep. After the period of 50 years the creep limit is already reached by the normal creep strain, so no effects of the mechano-sorptive creep are visible.

Besides that, the model according to Toratti [51] reaches the creep limit already for medium amplitudes ($\Delta RH > 6\%$), whereas in the models according to Mårtensson [38] and Becker [2], the creep coefficient still increases with an increasing amplitude of the relative humidity.

Considering the influence of the average relative humidity (see Fig. 4.6) the creep coefficient increases within the models according to Mårtensson [38] and Becker [2], whereas in the model according to Toratti [51] the creep coefficient decreases. The decrease of the creep coefficient is due to an increasing elastic deflection with an increasing moisture content and the dependence of the creep strain on a moisture independent Modulus of Elasticity (see $J_{u,ref}$ in Eq. (3.10)). The model according to Hanhijärvi [22] is hardly influenced, since the creep limit after 50 years is already reached.

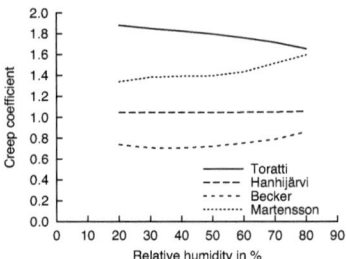

Figure 4.6: Influence of the average relative humidity on the creep coefficient

Since the moisture content influences the creep deformation, the thickness also influences the creep deformation, because the thicker the cross section, the lower the minimum moisture variation and therefore the lower the mechano-sorptive creep strain. In Fig. 4.7 the creep

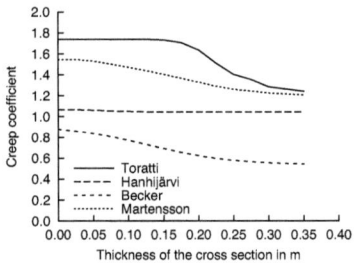

Figure 4.7: Influence of the thickness of the cross section on the creep coefficient

coefficient in dependence on the thickness of the cross section is given. As seen in Fig. 4.7 in the model according to Toratti [51] the creep limit is reached even for cross sections with thicknesses of up to 15cm to 20cm, whereas the model according to Hanhijärvi [22] is hardly influenced by the dimensions. In the other models, the creep coefficient is reduced by the thickness of the cross section up to 0.3.

Regarding the influence of the stress level (see Fig. 4.8), it becomes obvious that the mechano-sorptive creep in the models according to Toratti [51], Becker [2] and Mårtensson [38] has a linear influence on the temporal strains. So the creep coefficient is independent of the stress level as shown in Fig. 4.8. In the model according to Hanhijärvi [22] the stress level influences the creep coefficient. However, in this model the creep coefficient varies between 0.9 to 1.1 for the stress level in the serviceability limit state.

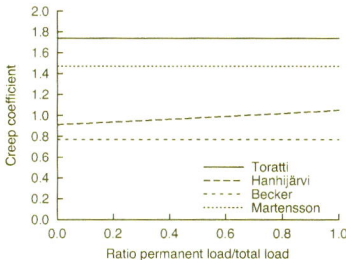

Figure 4.8: Influence of the stress level on the creep coefficient

4.4 Influence of the variability of the material properties on the creep coefficient

The purpose of the models is to describe the "real" behavior of timber beams. But normally they are verified on species without any defect. The influence of the variability of the material property is often considered by a scattering of the Modulus of Elasticity[1] (see Hartnack [24] and Becker [2]). For determining the influence of the variability of the Modulus of Elasticity a case study is performed.

As shown in Fig. 4.9 the Modulus of Elasticity hardly influences the creep coefficient except for the model according to Mårtensson [38]. The Modulus of Elasticity has a linear influence on the strain in most of the models. Therefore there is no influence on the creep coefficient in the model according to Toratti [51] and Becker [2], whereas in the model according to Hanhijärvi [22] a very slight influence can be determined since the creep coefficient depends on the stress level. In the model according to Mårtensson [38] the Modulus of Elasticity influences the creep coefficient remarkably.

For the evaluation of the influence of the variability of the Moduls of Elasticity on the effective creep coefficient, the Modulus of Elasticity is assumed to follow the statistical distribution of the Modulus of Elasticity according to [30] (Average Modulus of Elasticity equal $11500 N/mm^2$; standard deviation equal $2000 N/mm^2$). The variation of the Modulus of Elasticity is considered in each point of the structure independently. So there is no influence if the points belong to the same board in a glulam beam.

When considering the variability of the Modulus of elasticity in glulam beams, the creep coefficient in constant climate is about 0.49 (see Fig. 4.10); the standard deviation of the creep coefficient is 0.0016. Since the standard deviation is quite small, the influence of the variability of the Modulus of Elasticity in glulam beams can be even neglected in the model according to Mårtensson [38]. Therefore the influences "time-dependent behavior" and "variability of the material properties" may be split in glulam beams, which simplifies coupled problems such as buckling of columns or lateral torsional buckling.

---

[1] In the model according to Becker [2] the variability of the shrinkage/swelling coefficient is also considered. Nevertheless, no variability of this coefficient is considered here, since the shrinkage/swelling coefficient differs in the different models and the models are linked to the coefficient specificated in the models.

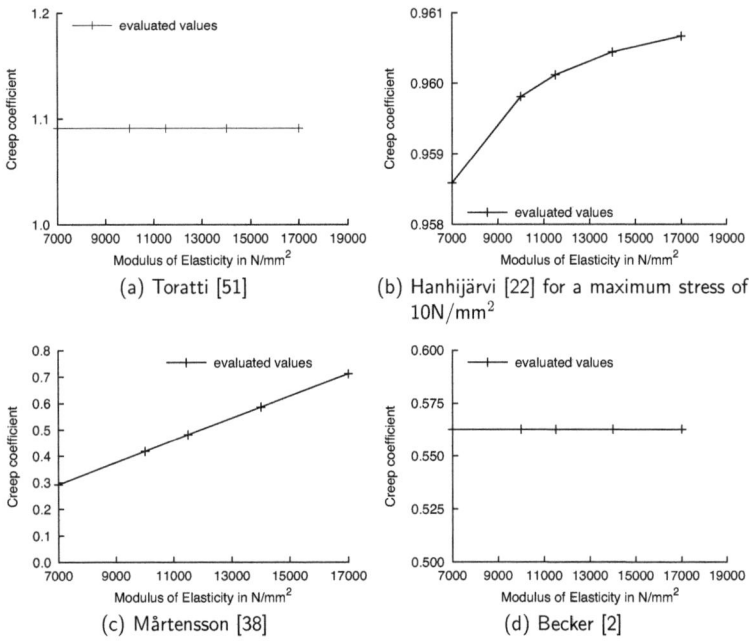

Figure 4.9: Influence of the Modulus of Elasticity on the creep coefficient

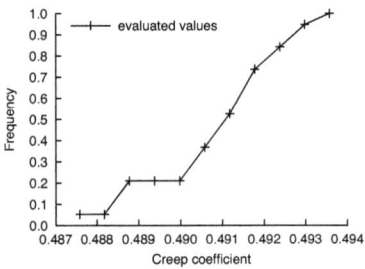

Figure 4.10: Influence of the variability of the Modulus of Elasticity on the creep coefficient according to Mårtensson [38]

4.5 Effects for the application in timber-concrete composite structures

When applying these models to timber-concrete composite structures, the different development of the creep strain leads to different results. Since timber creeps faster than the concrete in the model according to Becker [2], timber reduces its stresses. In the other models the concrete creeps faster, so the stresses in timber increase. This increase of the stresses due to the redistribution of the internal forces can amount to 20% of the elastic values and appears in the range of three to seven years after casting and loading of the timber-concrete composite slab, especially if shrinkage of the concrete is reduced (see Fragiacomo [15], [16] and [48]). So in the models according to Hanhijärvi [22], Mårtensson [38] and Toratti [51] the points in time of $t = 0$, $t = \infty$ and $t = 3$ to 7 years have to be considered, whereas in the model of Becker [2] only $t = 0$ and $t = \infty$ has to be taken into account. So the determi-

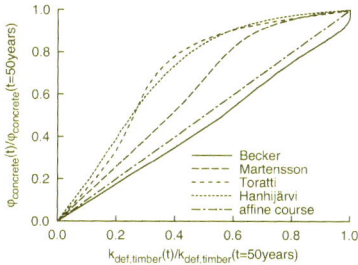

Figure 4.11: Development of the creep coefficient of concrete in dependence on the creep coefficient of timber for constant humidity

nation of the effective creep coefficients in timber-concrete composite structures, proposed in [48], is valid only for the model according to Hanhijärvi [22], which was the basis of the studies in [48].

4.6 Conclusions

In this chapter, the responses of the models applied to identical situations are compared. The obtained creep coefficients of the models vary strongly in constant climate as well as in varying climate in dependence on the model used. Therefore, it is impossible to recommend a model for analyzing the long-term behavior of structures.

In order to consider the variability of the material properties, the Modulus of Elasticity is assumed to be statistically distributed (see among other Hartnack [24] and [30]). The parameters of the models themselves, as e.g. compliance or retardation time, are not varied since the coefficients of variation of these parameters are hardly known. In three models there is no influence of the variability of the Modulus of Elasticity, whereas in one model the influence is small, that it can be neglected.

So only the variation of the Modulus of Elasticity needs to be considered in the evaluation of the elastic deformation, since the creep deformation depends linearly on the elastic deformation.

$$w_{\text{creep}}(MoE, CoV_{MoE}) = k_{def} \cdot w_{\text{elastic}}(MoE, CoV_{MoE}) \quad (4.2)$$

where $w_{\text{creep}}(MoE, CoV_{MoE})$ creep deformation in dependence on the variation of the Modulus of Elasticity
  $w_{\text{elastic}}(MoE, CoV_{MoE})$ elastic deformation in dependence on the variation of the Modulus of Elasticity
  k_{def} creep coefficient evaluated by the models

For predicting the long-term behavior of timber-concrete structures the different creep values after a period of time of 50 years as well as the different temporal development lead to quantitatively as well as qualitatively different conclusions for the design of this type of composite slabs (see among others Fragiacomo [15], Grosse et al. [18], Bou Saïd [3] and [48]).

The different models lead to too large differences in their responses. Therefore the models should be validated with test results. However, up-to-now this is hardly possible, since these tests result do not exist due to the long duration. Since the original purpose of these models is to show the creep deformations of real structures and not of test specimen, measurements of the creep deformations in real structures should validate one of these models.

5 Determination of the existing long-term deformations

5.1 General

As shown in Chap. 4, the responses of the different rheological models differ significantly in the prediction of the time-dependent behavior (see e.g. Fig. 4.7). For the same load history and the same surrounding conditions the creep coefficient can vary between 0.6 and 1.8. The large differences in the responses of the different models raise the question why the differences are so large and which model should be used for further studies on the time-dependent behavior of complete structures.

Models are developed by adapting functions to measured test results. Strictly spoken, the duration of the tests limits the period of the validity of the model. However, the duration of the tests is limited in length due to the possibility of performing long-term tests (see Fig. 5.1). As shown in Fig. 5.1 most of the tests, given in literature, last less than one year.

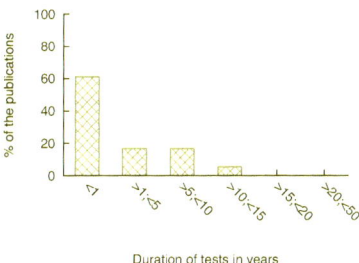

Figure 5.1: Duration of tests in dependence on the number of publications (extracted from the data given in Becker [2])

The maximum duration in the compilation given in Becker [2] is 12 years. When comparing the response of the models within the time range of one year, the determination of the deflections within the first year hardly differs from each other and from the tests as shown in Fig. 5.2.

All models are valid within the range of time, for which they have been validated. The differences in the studied models are caused by the extrapolation, since small differences in the inclination of the course of the creep strain at the end of the validated range of time lead to large differences at the assumed end of the anticipated lifetime of a building after $t = 50$ years.

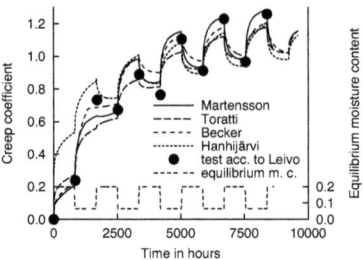

Figure 5.2: Comparison of the models to the test by Leivo [35]

The differences in the prediction of the time-dependent behavior are quite large, maybe too large for the design. Therefore the only way to validate a rheological model and to get values for the creep coefficient of timber beams after 50 years is to determine the creep coefficient by experiment. But tests can hardly be performed due to the duration of these tests. However, a lot of buildings have been built using timber as a structural material. So it can be asked, whether these buildings can be used for the backward-determination of the creep coefficients.

5.2 Determination of creep coefficient by measurements

5.2.1 Procedure

In the following, creep coefficients are identified by measurements on real structures. In principle, there are two different ways to approach the problem:
- performing an elastic calculation and measuring the current deformations
- performing a test loading and measuring the initial deformations and the deformations under load.

Regarding the determined elastic and the measured deformation, in principle, the creep coefficients can be determined by

$$\varphi = \frac{w_{\text{measured}} - w_{\text{elastic}}}{w_{\text{elastic}}} \tag{5.1}$$

With this procedure effective creep coefficients can be determined. These effective creep coefficients represent the creep coefficient, which should have been used by the engineer, in order to predict the deformation after 50 years.

This effective creep coefficient differs from the real material creep coefficient due to e.g. the neglect of the real load history and the neglect of the difference of swelling under compression or tension. Therefore, the effective creep coefficient will be labeled as φ in the following. Besides that, only members subjected to bending will be measured for the evaluation of effective creep coefficients, since the deformation due to bending is easier to measure compared to e.g. the deformation of columns.

5.2.2 How to deal with the unknown influences?

This type of measurements bears several uncertainties, which may influence the results. However, it should be possible to minimize the influences of the unknown parameters by a specified selection of elements.

- **initial moisture content**: For solid timber beams, Carstensen [5] recommends to assume a moisture content of 20% to 30%, since the mechanical properties hardly change with a moisture content above the fibre saturation point. Since this is quite a wide range of possible moisture contents, the influence of different moisture contents on the deformation after 50 years is studied by using the model according to Toratti [51]. Fig. 5.3 shows that the influence of the moisture content is less than 10% of the final deformation for the study cross section of 10cm × 10cm.

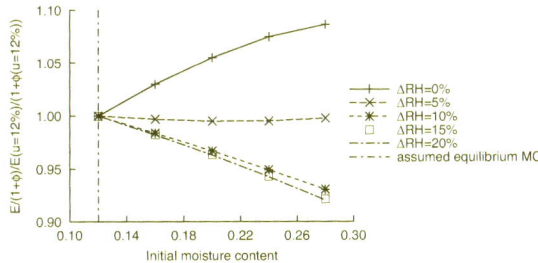

Figure 5.3: Influence of the initial moisture content on the effective bending stiffness after 50 years according to Toratti [51]'s model in dependence on the amplitude of the annual changing relative humidity ΔRH (b × h = 10cm × 10cm)

These relatively small influences of the initial moisture content are mainly caused by the elastic recovery and the different eigenstresses due to the different initial moisture contents and the resulting drying process. Most of the rheological models as given in Becker [2], Mårtensson [38] or Toratti [51] link the changing moisture content with the stresses by following principle equation

$$\varepsilon_{ms} = \int f \left(\int \hat{h} \cdot |du| \right) d\sigma \\ = \int f \left(\hat{h} \cdot \sum |du| \right) d\sigma \tag{5.2}$$

where ε_{ms} mechano-sorptive creep strain
 σ stress
 $|du|$ absolute moisture variation
 $\hat{h}$ parameter

If constant stresses are assumed and therefore the eigenstresses caused by the drying/wetting process are neglected, the mechano-sorptive creep is a function of the accumulation of the moisture variations in the cross section.

In Fig. 5.4 the accumulated moisture content $\sum |du|$ of a cross section (h × b = 20cm × 10cm) with an initial moisture content of 12% is compared to the accumulated moisture content of the same cross section with an initial moisture content of 30% in a climate with an average relative humidity RH_{av} of 65% and an annual amplitude ΔRH of 15%. As shown in Fig. 5.4 the accumulated moisture content hardly differs

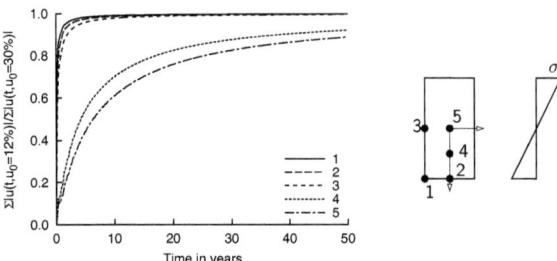

Figure 5.4: Comparison of the accumulated moisture content in a cross section of h × b = 20cm × 10cm with an initial moisture content of 12% related to the same cross section with an initial moisture content of 30% in changing climate ($RH_{\text{average}} = 65\%, \Delta RH = 15\%$)

in the outer layers after 5 to 10 years, whereas in the centroid of the cross section differences in the accumulated moisture content still exist. The differences in the outer layers between an initial moisture content of 12% related to the initial moisture content of 30% are small or after 10 years even negligible, since the moisture variations due to the changing climate dominate the accumulated moisture.

In order to determine the influence of the different accumulation of the moisture content, the curvature neglecting the eigenstresses can be approximated by

$$\kappa_{ms} = \int\int g\left(\int \hat{h} \cdot |du|\right) d\sigma dA \stackrel{\substack{\text{assuming}\\ \text{constant stress}}}{=} \int g\left(\int \hat{h} \cdot |du|\right) \sigma(z) dA$$
$$= \int g\left(\int \hat{h} \cdot |du|\right) \sigma_{max} \cdot \frac{2 \cdot z}{h} dA = \frac{M}{EJ} \cdot \underbrace{\int f\left(\int \hat{h} \cdot |du|\right) \cdot \frac{2 \cdot z^2}{h} dA}_{\varphi_{eff}} \quad (5.3)$$

where A area of the cross section
       u moisture variation
       σ stresses
       α coefficient of shrinkage/swelling
       $\hat{h}$ parameter
       f,g function
       h height of the cross section
       M bending moment
       EJ bending stiffness
       φ_{eff} effective creep coefficient

If constant stresses – i.e. neglect of eigenstresses due to different shrinkage or swelling and the neglect of the dependence of the shrinkage/swelling-coefficient α on the stress

– are assumed, the ratio of the effective creep coefficient in cross sections with different initial moisture contents can be determined. As shown in Fig. 5.5, the effective

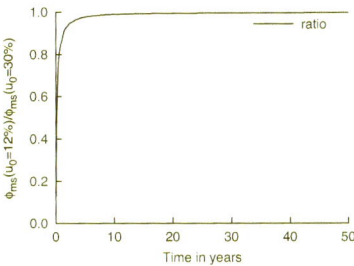

Figure 5.5: Ratio of the effective creep coefficient caused by mechano-sorptive creep for an initial moisture content of 12% and an initial moisture content of 30%

creep coefficient due to mechano-sorptive creep is nearly independent of the initial moisture content. The reason is, that in the ranges with significant differences in the accumulated moisture the stresses are low in elements subjected to bending and the mechano-sorptive creep links the accumulation of the moisture to the stresses. Therefore, the differences in the moisture accumulation in the inner layers do not have a big influence on the effective creep coefficient. For this reason, the initial moisture content should not influence the results significantly, especially if elements with small widths are measured, since it is expected, that the eigenstresses in these elements are quite small.

- **imperfections:** Since the initial state of the elements is unknown, the imperfection can influence the creep coefficients, derived from measurements. However, the variation of the imperfection can be described by the following equation given by Hartnack [24].

$$e_0 = (1.150 \pm 10.538) \cdot 10^{-4} \cdot L \qquad (5.4)$$

where L span of the element
e_0 imperfection of the element

The influence of the imperfection can be reduced, if several identical elements in the same building are measured. If a sufficient number of elements are measured, the average imperfection of $e_{0.mean} = 1.150 \cdot 10^{-4} \cdot L$ can be assumed for the average deformation. In order to determine this sufficient number of identical elements, a Monte-Carlo-simulation with the variation of the imperfection according to Eq. (5.4) is performed, yielding the number of measurements required for a certain accuracy (see Fig. 5.6). As shown in Fig. 5.6, the influence of the imperfection is less than 5% on average, if more than 10 measurements are performed. If in addition to this, elements are used, where only wider limits concerning the deformations are required, the influence of the imperfection decreases additionally (see Fig. 5.6(b)).

In the investigations of this study barns, garages or machine shops for agricultural use will be used for the measurements of the existing deflection, since DIN 1052 [8] only defines the limit of the deflection of about $L/150$ for these types of buidlings.

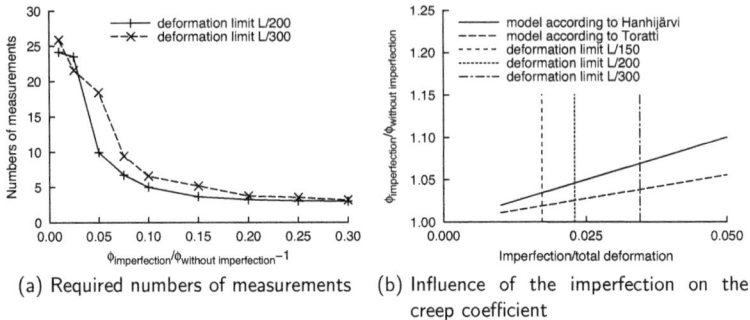

(a) Required numbers of measurements

(b) Influence of the imperfection on the creep coefficient

Figure 5.6: Influence of the imperfection on the creep coefficient

Besides that only elements are considered, for which the deformation is affine to the assumed elastic deformation. Applied to the measurements, only these elements are considered, where the deformation in the quarter of the beam length is in the range of affine deformation ± 50% of the mid-span value (see Fig. 5.7). It has to be mentioned,

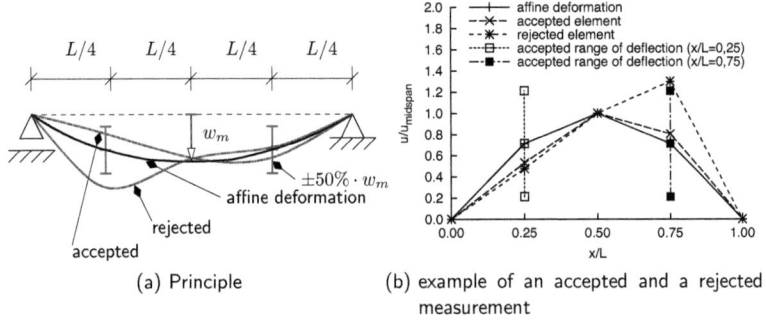

(a) Principle

(b) example of an accepted and a rejected measurement

Figure 5.7: Definition of an accepted affine deformation

that this boundary of ± 50% is chosen as a reasonable but arbitrary condition.

- **course of the temperature and the relative humidity:** The average temperature and the relative humidity respectively, as well as the variations of those influence the time-dependent behavior in a strong manner (see among others Toratti [51], Hanhijärvi [22], Becker [2], Mårtensson [38], Gressel [17]).

The drying process is quite a slow process; shown in the fact, that the drying of timber in natural climate takes 2 to 3 years or in the measured course of the moisture given in Hanhijärvi [22] (see Fig. 5.8). In Hanhijärvi [22] a spruce beam with the dimensions of breadth × height = 15mm x 75mm and an initial moisture content of 20% needs 6 days to achieve the moisture content of about 10% if the relative humidity suddenly drops from 90% to 30% and the moisture can evacuate at all four sides of the cross section. Therefore only larger cycles of the relative humidity should influence the

5.2 Determination of creep coefficient by measurements

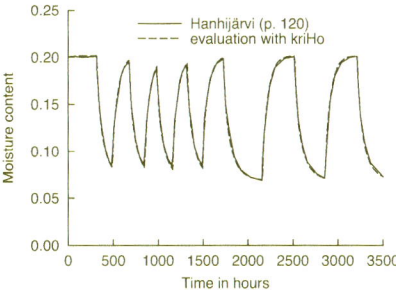

Figure 5.8: Course of the moisture content of a test specimen with a thickness of 2mm during the changing of the relative humidity between 30% and 90% (see Hanhijärvi [22])

moisture content in the inner layers and the mechano-sorptive creep strain of the cross section.

For the approximation of the temporal course of the relative humidity, Häglund [21] proposes to model the real course of the relative humidity by an annual and a daily variation. In Fig. 5.9 the maximum and minimum moisture content in dependence on the annual amplitude of the relative humidity based on evaluations according to the proposal given in Toratti [51] is shown. As can be seen in this figure, only the annual variation seems to be of importance, since the daily cycle only influences the moisture content in the outer layers of the cross section.

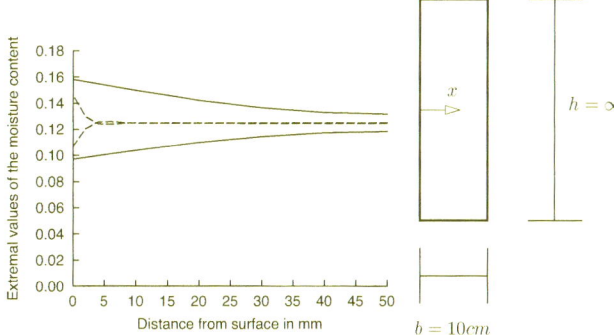

Figure 5.9: Maximum and minimum moisture contents of a cross section with 10cm width in dependence on the duration of the variation of the relative humidity according to the model given in Toratti [51] ($RH_{\text{average}} = 50\%$, $\Delta RH = 20\%$)

The moisture variations lead to mechano-sorptive creep and therefore to larger creep coefficients in the outer layers. In Fig. 5.10 the influence of the daily moisture variations for different ratios between the creep coefficient in the center of the beam and an assumed increased creep coefficient within the outer 4mm of the cross section is shown. The decrease of the effective bending stiffness due to daily variations is less than 10%,

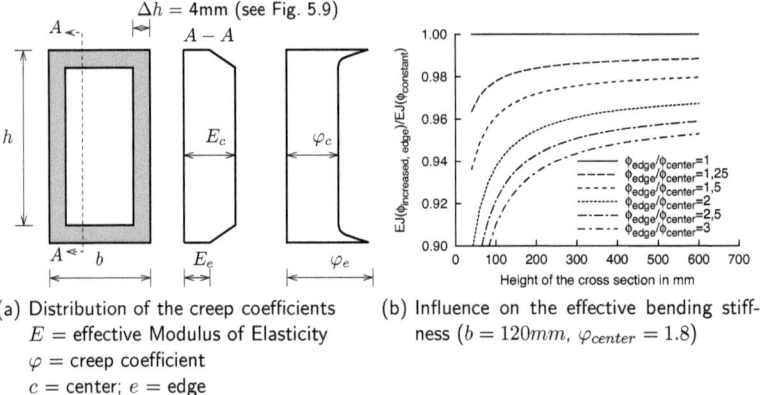

(a) Distribution of the creep coefficients
E = effective Modulus of Elasticity
φ = creep coefficient
c = center; e = edge

(b) Influence on the effective bending stiffness ($b = 120mm$, $\varphi_{center} = 1.8$)

Figure 5.10: Influence of the increased creep due to daily moisture variations on the effective bending stiffness

even if the cross section has a dimension of 100mm × 120mm and the creep coefficient due to the increased moisture variation is 3 times larger than the creep coefficient in the ranges, which are not influenced by the daily moisture variations.

Since the natural climate is mainly influenced by the annual and the daily variations (see Häglund [21]) and the daily variations can be neglected, it is assumed, that the data given by local weather stations are sufficient for describing the local climate in the building and therefore in the cross section, provided the building is open and not heated, but protected against direct wetting. So the elements fulfill the requirements of service class 2 according to DIN 1052 [9].

- **shrinkage of timber**: If shrinkage is defined as stressless strains, caused by drying and the influence of the mechanical strain on the coefficient of shrinkage is not considered in difference to the rheological models (see among others Eq. (3.13) and Eq. (3.23)), the influence of the shrinkage can be neglected in those cases, when the moisture variations can penetrate the cross section over all four surfaces. So it is assumed, that differences in the moisture content related to the symmetry axis of the cross sections hardly appear. In this case, shrinkage should not cause any curvature, so shrinkage has no influence on the measurements.

However, is it justified to neglect the dependence of the shrinkage coefficient on the mechanical strain? In this study, the creep coefficient in bending for practical use is determined and not the real material properties. For this reason, it seems sufficient to integrate the dependence of the coefficient of shrinkage on the mechanical strain into the effective creep coefficient although shrinkage is not directly linked to creep.

- **load history**: Since the load history influences the creep deformations, it should be reconstructed over the whole time. However, for slabs it is hardly possible, since the load history can often not be defined, whereas roof structures are mainly loaded by dead load and snow loads, which can be approximated by the data of weather stations for the past years.

5.2 Determination of creep coefficient by measurements

DIN 1055 [10] already proposes rules for the consideration of snow load in the long-term. According to this rule the influences of the snow load on the creep deformation can be neglected, if the altitude above sea level of the building is less than 1000m. Using this rule, creep coefficients can be determined, which fit within the scope of the valid standards for the practical use.

- **structural system**: For the determination of the creep coefficient primarily single span girders are measured. Performing these measurements, relative deformations are monitored (see Fig. 5.11). With these 5 measured deflections, the course of the deflection can be evaluated. Therefore the influence of unknown clamping at the support or the complete structural system can be checked, especially, if the test load is compared to the expected course of the deflection according to an elastic calculation.

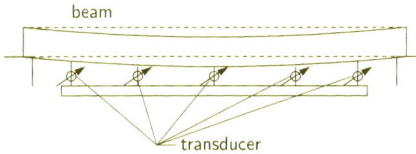

Figure 5.11: Performing of the measurements

- **material properties**: In Carstensen [5], "only" an effective Modulus of Elasticity of the single cross sections is determined, which should have been used for the structural design.

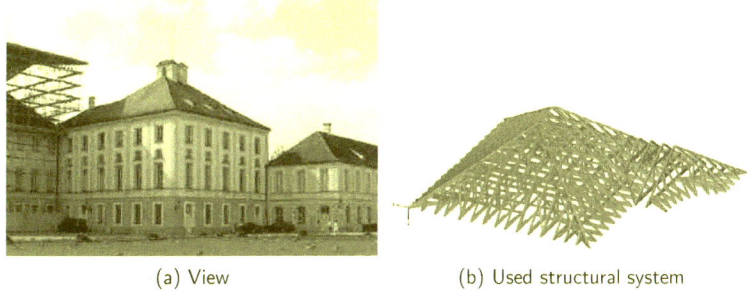

(a) View   (b) Used structural system

Figure 5.12: Roof structure of the chapel of the "Schloss Nymphenburg", Munich (see Dieringer [7])

In Dieringer [7] and Lühr [36] the creep coefficient is determined according to the following procedure (see Fig. 5.12 and Fig. 5.13)

– visual grading of timber according to DIN 4074 [12]
– evaluation of the elastic deflection based on the Modulus of Elasticity of the graded strength class
– measurement of the existing deflection

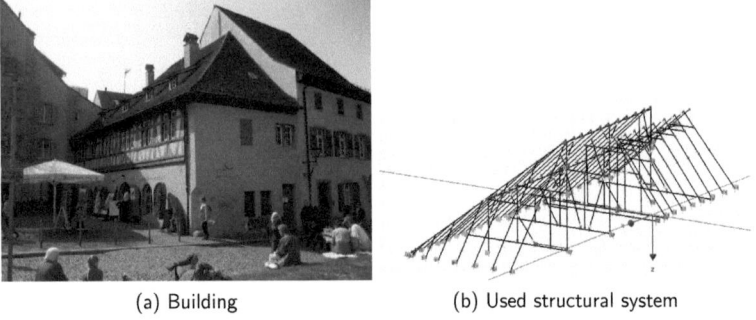

(a) Building   (b) Used structural system

Figure 5.13: Roof structure of the Alte Münsterbauhütte, Freiburg i. Br. (see Lühr [36])

– determination of the effective creep coefficient

Real material properties are not determined, since e.g. the variability of the Modulus of Elasticity is considered in the creep coefficient.

This procedure can be accepted, since creep coefficients are to be determined, which lead to the correct calculated deformation for the design after the used period of time.

On the other hand, the procedure mentioned above leads to larger variations in the values. As the rheological models should be verified by comparing the results to the evaluated deformations, the influence of the material properties should be reduced as much as possible.

For the determination of the creep coefficient, the elastic deflection is of interest. In principle, it can be determined by unloading the system. However, this is hardly possible in a real structural system. Another possibility is to determine the integral elastic stiffness of the system by a test loading and measuring the additional deflection. Based on these results, the Modulus of Elasticity can be evaluated and the timber grade of the element according to DIN 4074 [12] can be estimated. Additionally, the moisture content is determined. So the existing material properties and therefore the elastic deformation can be estimated.

As shown, several uncertainties can influence the results. Nevertheless, if the elements are selected properly, it should be possible to determine effective creep coefficients, which should have been used in the structural design in order to predict the real deformation after a period of time of 50 years.

In order to perform these measurements, several requirements for the building are defined in order to reduce the influences of the unknown uncertainties:

- free accessibility to the structure
- several identical elements
- low requirements on the deflection
- year of construction between 1950 and 1980
- "defined" load history, e.g. the only variable load is the snow load

- opened and protected, but not heated structures

Summing up all these requirements, roof structures of agricultural machine shops, barns, garages or agricultural storehouses are suited best for the measuring of the existing deflection and to determine the effective creep coefficient.

5.3 Determination of effective creep coefficients in the region of Tübingen, South-West Germany

For the determination of creep coefficients by measurements, access to the buildings is necessary. Most of the buildings are in use and therefore unfortunately the willingness of the owner to allow the desired measurements is often limited. The region of Tübingen is chosen for reasons of personal relations to the owners of the buildings. This region is located in the center of the federal state of Baden-Württemberg, South-West Germany (see Fig. 5.14).

Figure 5.14: Location of the region of Tübingen (see wikipedia.de [54] and maps.google.de [37])

Since the changing moisture content influences the creep strain, the relative humidity as well as the snow load should be known. Unfortunately, no weather data directly from Tübingen are available. For this reason, the data of Stuttgart airport in Leinfelden-Echterdingen, 23km away from Tübingen, are used to characterize the surrounding conditions.

The reconstructed climate is given since January 01, 1953 in Fig. 5.15(a), based on the data given in [53]. In Fig. 5.15(b) the results of a frequency analysis are shown. As can be seen, the variation of the relative humidity is mainly caused by a daily and an annual humidity variation.

For the snow load a similar database is available. However, the average snow height is 5cm for 30 days per year. Additionally, the altitude of the measured buildings in the region of Tübingen is less than 1000m above sea level, so snow should not lead to creep deformations according to DIN 1055 [10]. Due to the short time and also according to the rules in DIN 1055 [10] the snow load is neglected.

For the determination of the effective creep coefficient after 50 years the rafters in 4 buildings with 56 rafters in total and an age between 45 and 57 years have been measured (see Fig. 5.16). 32 of the 56 rafters are round wood beams with an diameter between 7.5cm and 12cm, where often the bark still exists. The other 24 elements are sawn timber with

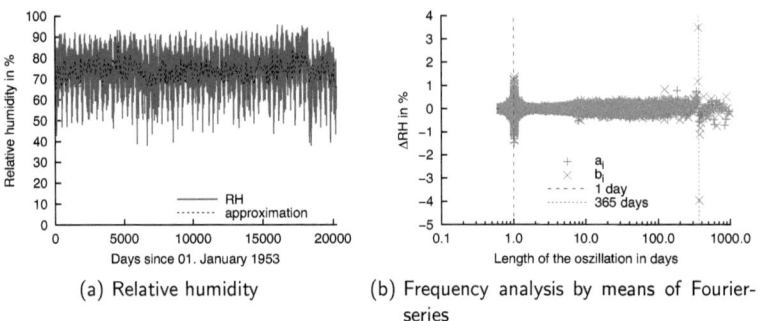

(a) Relative humidity (b) Frequency analysis by means of Fourier-series

Figure 5.15: Relative humidity in Leinfelden-Echterdingen

breadth × height = 5.7cm × 11.7cm. The structural material in all buildings is identified as softwood.

Since only elements with an affine existing deflections are considered in the evaluation of the effective creep coefficient, 20 elements are sorted out (see Tab. 5.1). In Fig. 5.17 the

Table 5.1: Used elements for the determination of the creep coefficient in the region of Tübingen

Duration of load:	45 years to 57 years
Elements:	sawn timber & round wood
• sawn timber	19 accepted rafters; 57mm × 117mm
• round wood	17 accepted rafters; ∅≈ 75mm to 120mm

statistical distribution of the remaining 36 elements is shown. The average value and the standard deviation is given in Tab. 5.2.

Table 5.2: Average values and standard deviations of the creep coefficient of Tübingen

	average value	standard deviation	coefficient of variation
sawn timber	2.23	1.09	0.49
round wood	2.23	0.87	0.48
all elements	2.23	0.97	0.48

As shown in Fig. 5.17 as well as in Tab. 5.2, a large variability of the creep coefficients in timber exists. So the coefficient of variation for all elements is 0.48.

The correlation ρ between the measured values and the normal distribution is determined by

$$\rho = \frac{1}{n} \cdot \frac{\sum_{i=1}^{n} (x - \mu_x) \cdot (y - \mu_y)}{\sigma_x \cdot \sigma_y} \tag{5.5}$$

5.3 Determination of effective creep coefficients in the region of Tübingen, South-West Germany

(a) Jungviehweide, Ofterdingen;
456m above sea level, round wood,
48°24'54.76"N, 9°01'10.88"E

(b) Marienstr. 1, Ofterdingen;
429m above sea level, round wood,
48°24'55.98"N, 9°01'58.70"E

(c) Marienstr. 1, Ofterdingen;
429m above sea level, round wood,
48°24'56.29"N, 9°01'58.97"E

(d) Mörike-school, Tübingen;
328m above sea level, sawn timber,
48°30'23.12"N, 9°03'21.00"E

Figure 5.16: Buildings, where the measurements are performed

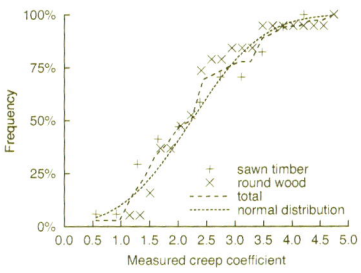

Figure 5.17: Distribution of the creep coefficient

where n numbers of measurements
x measured values
y value evaluated by the normal distribution
μ average value
σ standard deviation

The correlation between the normal distribution and the measured values ρ of 0.94 is determined. Therefore the distribution of the creep coefficients may be assumed to be normal distributed.

In Fig. 5.18 the measured creep coefficient is set into relation to the Modulus of Elasticity evaluated by using the elastic deformations of the test loading. As shown, no significant

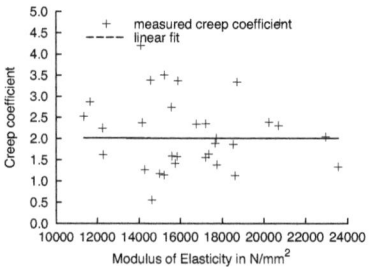

Figure 5.18: Creep coefficient in dependence on the determined Modulus of Elasticity

influence of the Modulus of Elasticity can be found within the measured set. Therefore the variability of the creep coefficients seems to be independent of the Modulus of Elasticity. However, it has to be noticed, that the boundary conditions, such as clamping at the support or increased stiffness due to the load transfer to the end wall by means of the battens (see Fig. 5.19), are included in the Modulus of Elasticity. When measuring the elastic deformation

(a) Clamping at the support due to a large groove depth for the connection between rafter and purlin

(b) Load transfer to the end wall by means of the battens

Figure 5.19: Influences on the evaluated Modulus of Elasticity

for evaluating the Modulus of Elasticity irrespective of the clamping, a large Modulus of Elasticity can be evaluated. So the independence of the creep coefficient on the Modulus of Elasticity and the variability of the creep coefficient as material immanent variability can not be fully verified.

Apart from the large variability of the values, no evident difference between the time-dependent behavior of the sawn timber and round wood may be found. The average creep coefficient as well as the standard deviation of the set with pure sawn timber and the set of round wood match quite well.

Since there is still bark on the round wood elements, it is likely, that the initial moisture content was much higher than the initial moisture content of the sawn timber. Therefore the initial moisture content hardly seems to influence the creep deformation after 50 years, since the changing surrounding conditions determines the moisture dependent creep (= mechano-sorptive creep, see Sec. 5.2.2).

According to DIN 1052 [9] and Eurocode 5 [14] the situations in the buildings can be classified as service class 2 - conditions (protected, but not heated), the creep coefficient recommended by these standards for this service class is 0.8. Following the distribution given in Fig. 5.17 and in Tab. 5.2 the current standard provides the 7% fractile value.

Therefore the creep coefficient according to DIN 1052 [9] and Eurocode 5 [14] is too small for determining average values of the deformations, at least for the measured rafters under the given conditions.

5.4 Comparison of the measured creep coefficients to the responses of the models, to common rules in the standards and to studies from literature

In Fig. 5.20 the average measured creep coefficients are compared to the evaluation by the rheological models and the standards DIN 1052 [9] and Eurocode 5 [14] respectively, assuming an initial moisture content below the fibre saturation point. As can be seen in

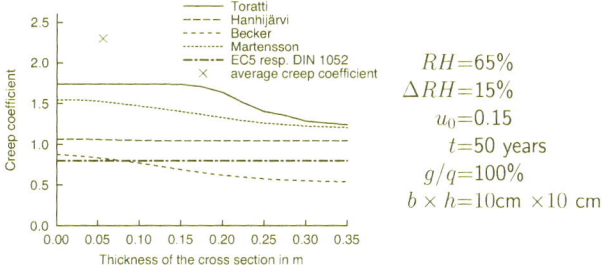

Figure 5.20: Influence of the thickness of the cross section

Fig. 5.20, all models underestimate the creep coefficients derived by measurements. However, the model according to Toratti [51] shows the lowest differences between the measured and the evaluated values.

The values given in the standards underestimate the creep coefficient. However, Eurocode 5 [14] and DIN 1052 [8] regulate, that the creep coefficient has to be increased about 1.0 if the initial moisture content is in the range of the fibre saturation point. For these cases the creep coefficient in service class 2 becomes 1.8.

The difference between normal conditioned timber and timber with a moisture content in the range of the fibre saturation point is, that the wetter timber dries more strongly than the normal conditioned timber. This should lead to a stronger mechano-sorptive creep, since the mechano-sorptive creep is often regarded as a function of the accumulation of the absolute moisture variation.

$$\varphi_{\text{mechano sorption}} = f\left(\sum |\Delta u|\right) \tag{5.6}$$

If the accumulation of the absolute moisture content of a cross section in a climate with an amplitude of the relative humidity of 15% is determined for different initial moisture contents, only very small differences arise, especially in the outer layers of the cross section (see Fig. 5.4). Therefore, the creep coefficient should be equal to the creep coefficient of a cross section with a moisture content in the range of the fibre saturation point, if the modeling of the mechano-sorptive creep as a function of the accumulated moisture is valid as in most of the models. In this case the creep coefficient is

$$\varphi = \underbrace{0.8}_{\substack{\text{creep coefficient of service-}\\\text{class 2}}} + \underbrace{1}_{\substack{\text{increase due to a moisture}\\\text{content in the range of the fi-}\\\text{bre saturation point}}} = 1.8 \tag{5.7}$$

This larger creep coefficient compared to the rules in Eurocode 5 [14] or DIN 1052 [9] is confirmed by the studies of Rautenstrauch [46] and Moorkamp [40]. According to these studies the creep coefficient in natural climate can be determined according to the following equations given by Rautenstrauch [46]

$$\varphi = 0.0292 \cdot t^{0.292} + 0.0907 \cdot \Delta u - 0.00237 \cdot \Delta u^2 \tag{5.8}$$

where t times in hours
       Δu moisture change in % due to drying of the timber from the initial moisture content to the equilibrium moisture content

and by Moorkamp [40] respectively

$$\varphi = 0.0598 \cdot t^{0.262} \tag{5.9}$$

where t times in hours

If both equations are extrapolated for a period of 50 years, creep coefficients in the range between 1.8 and 2.2 are evaluated (see Tab. 5.3).

Since the creep coefficient is the ratio between creep and the initial deflection, the initial deformation should be related to the initial moisture content, i.e. to the Modulus of Elasticity of the initial moisture content.

$$\varphi = \frac{w_{\text{creep}}}{w_{\text{elastic}}(u(t=0), t=0)} \tag{5.10}$$

5.4 Comparison of the measured creep coefficients to the responses of the models, to common rules in the standards and to studies from literature

Table 5.3: Creep coefficient according to Rautenstrauch [46] and Moorkamp [40] without elastic recovery

	creep coefficient
Rautenstrauch [46] ($\Delta u = 7\%$)	1.81
Rautenstrauch [46] ($\Delta u = 12\%$)	2.04
Rautenstrauch [46] ($\Delta u = 20\%$)	2.16
Moorkamp [40]	1.8

where w_creep creep deformation
$w_\text{elastic}(u(t=0), t=0)$ elastic deformation, considering the initial moisture content

In contrast to the studies in the laboratory as performed by Rautenstrauch [46], the initial moisture content cannot be determined in the measurements of the existing deflection, since the elements have reached their equilibrium moisture content at the point in time of the measurement. Therefore the results of the models according to Rautenstrauch [46] have to be transferred to the equilibrium moisture content by the following equation

$$\varphi = \frac{w_{cr}}{w_{el}(u = u^\star)} = \varphi_\text{Eq. (5.8)} \cdot \frac{w_{el}(u = u^\star)}{w_{el}(u = u_\text{eq.})} = \varphi_\text{Eq. (5.8)} \cdot \frac{E(u = u_\text{eq.})}{E(u = u^\star)} \quad (5.11)$$

where $\varphi_\text{Eq. (5.8)}$ creep coefficient according to Eq. (5.8)
       $E(u)$ Modulus of Elasticity
       w deformation
       u_{eq} equilibirum moisture content
       $u^\star$ initial moisture content, resp. moisture content, the creep coefficient is related to

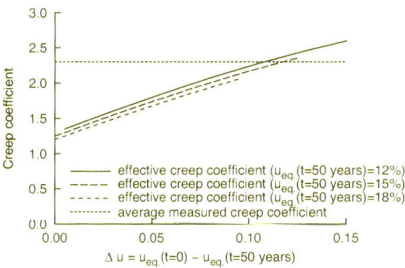

Figure 5.21: Creep coefficients based on Rautenstrauch [46] (see Moorkamp [40])

In Fig. 5.21 the creep coefficients according to Rautenstrauch [46] are related to the equilibrium moisture content after 50 years. As shown, the measured creep coefficient (see Tab. 5.2) is in the range of the creep coefficient determined by Rautenstrauch [46], if the difference between initial moisture content and equilibrium moisture content is in the range of 12%.

The same is assumed for the values given in the standards. If the relations for the Modulus of Elasticity according to Toratti [51] is used and the relation between to Modulus of Elasticity of 12% and of 30% as a moisture value in the range of the fibre saturation point is used, a coefficient of

$$k_{(u/u_0)} = 1.27 \qquad (5.12)$$

can be determined. This means, that the initial deflection of the cross section with an initial moisture content of 30% is 27% larger than the deflection of the same cross section with an initial moisture content of 12%. Therefore the creep coefficient is increased by 27%.

$$k_{def} = \frac{w_{creep}}{w_{elastic}} \qquad (5.13)$$

This results, in an effective creep coefficient of

$$\varphi = 1.27 \cdot \underbrace{1.8}_{\text{(see Eq. (5.7))}} = 2.3 \qquad (5.14)$$

Therefore, the measured creep coefficient is in the range of the values given in the standards, by Rautenstrauch [46] and by Moorkamp [40], respectively. So the possible inaccuracies of the measure procedure do not influence the measurements in an unacceptable way.

5.5 Summary

As shown the creep coefficients determined by measurements exceed the results of the models as well as the expected deformation in the standards. However, results from literature and an alternative interpretation of the standards lead to creep coefficients which are comparable to the creep coefficients determined by measurements. So the differences should not be caused by the procedure of the determination of the creep coefficients.

Therefore one of the studied model should be modified that way, that the creep coefficients after 50 years are comparable to those of the measurements. At the same time the modification should hardly change the responses within the first 5 to 10 years, since within this period the models are already validated.

6 Modified rheological model for the re-evaluation of the creep coefficients determined by measurements

6.1 General

In Chap. 4 different models have been compared to each other. Out of these studies, the model according to Toratti [51] shows the lowest differences between the measured and the creep deformation evaluated based on the measurements of Chap. 5. However, differences between the model and the values derived from measurements exist. For this reason, the model will be modified in order to describe the time-dependent behavior of timber within the region of Tübingen.

Concerning the modification it can be asked, whether these differences are caused mainly by mechano-sorptive creep or by normal creep.

Since in the models the accumulated moisture content determines the mechano-sorptive creep, the accumulated moisture content in the measured elements of the Mörike-school (see Fig. 5.16) is evaluated, using the parameters given in Toratti [51] (see Fig. 6.1). As

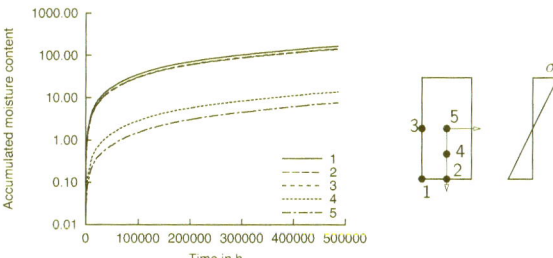

Figure 6.1: Accumulated moisture content for the elements at the Mörike-school (see Fig. 5.16)

can be seen, the accumulation of the moisture content is quite large. Tests performed by Mohager [39] seem to indicate that model B according to Toratti [51] underestimates the creep strain, especially when larger moisture variations and larger moisture accumulation take place (see Fig. 6.2(a)), whereas model D provides a sufficient accuracy between test and evaluation. If the moisture accumulation of the region of Tübingen is compared to the

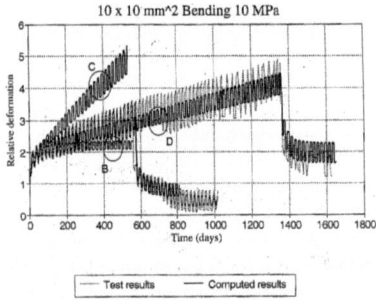 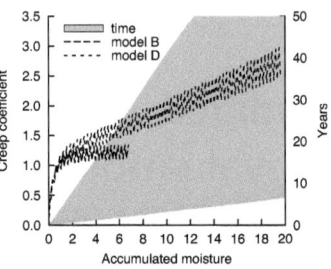

(a) Comparison of tests performed by Mohager [39] and the models B, D and C according to Toratti [51] (taken from Toratti [51])

(b) Creep coefficients of the creep test by Mohager [39] evaluated by the model B and D according to Toratti [51] in dependence on the accumulated moisture; time to reach the moisture accumulation in the elements of the Mörike-school

Figure 6.2: Influence of the accumulated moisture in the models B and D according to Toratti [51]

accumulation in the modeling of tests according to Mohager [39] (see Fig. 6.2(a)), model D instead of model B should be used, since the moisture accumulation in the measured elements is larger than in the tests (see Fig. 6.2(b)).

If model D according to Toratti [51] is used for the evaluation of the time-dependent deflection for the measured elements, the average deflection is overestimated (see Fig. 6.3).

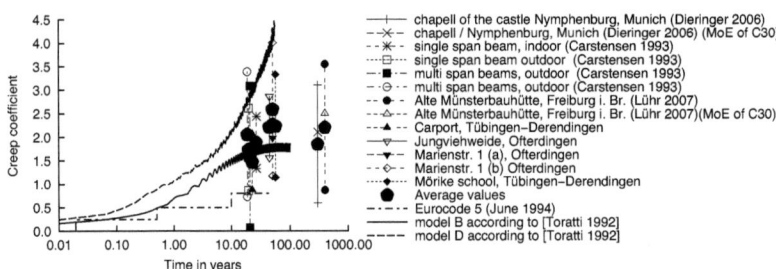

Figure 6.3: Comparison of the measured creep deformation with the evaluation based on the model B and D according to Toratti [51]

Model D describes the time-dependent behavior of the tests performed by Mohager [39] in a sufficiently exact manner, but it leads to larger differences in the prediction of the time-dependent deformations of the measured elements, although the accumulated moisture in the cross sections is comparable (see Fig. 6.2(b)). So model D fits to the measurements by Mohager [39] but not to the results of the measurements.

6.1 General

One reason for this discrepancy might be the dependence of the creep strain on the stresses. The nominal stresses of the tested beams are in the range of $10\,\text{N}/\text{mm}^2$. If the moisture variations are applied to the tests, the stresses at the outer layers vary between $-16\,\text{N}/\text{mm}^2$ to $-4.5\,\text{N}/\text{mm}^2$ and $14\,\text{N}/\text{mm}^2$ to $3.8\,\text{N}/\text{mm}^2$ respectively (see Fig. 6.4(a)). Due to these large

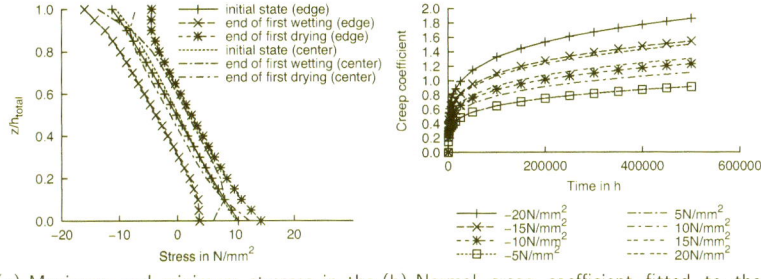

(a) Maximum and minimum stresses in the evaluation of the test according to Mohager [39] using the model B according to Toratti [51]

(b) Normal creep coefficient fitted to the model according to Hanhijärvi [22] in dependence on time and stress level

Figure 6.4: Stress distribution in the tests according to Mohager [39] and stress-dependent creep coefficient as possible reasons for differences between the model B according to Toratti [51] and the tests according to Mohager [39]

differences in the stresses during wetting and drying one reason for the differences between evaluation and test could be the influence of the stress level on the creep coefficient (see Fig. 6.4(b)). Since in the real structure the moisture variations are slower than those in the tests and the stress level is normally lower than in the tests, the eigenstresses are smaller in the real structure, since creeping leads to a reduction of the eigenstresses ($=$ relaxation).

Muszynski et al. [42] have measured the mechano-sorptive creep strain directly. Fig. 6.5 shows, that the influence of the moisture accumulation on the mechano-sorptive creep strain tends to some kind of creep limit. Nevertheless, Muszynski et al. [42] describe the mechano-

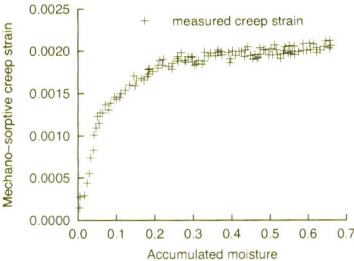

Figure 6.5: Measured mechano-sorptive creep strain in dependence on the accumulated moisture (see Muszynski et al. [42])

sorptive creep by a Kelvin-Voigt-body and a constant slope.

$$J = \frac{1}{E_{MS}} \cdot \left(1 - e^{-\nu \cdot m_1 \cdot E_{MS}}\right) + m_2 \cdot \nu \qquad (6.1)$$

where ν accumulated moisture content
$\sum |\Delta u|$
m_1, m_2 parameters given in Tab. 6.1
E_{MS} effective stiffness given in Tab. 6.1

Table 6.1: Parameters for the description of the mechano-sorptive creep used in the model according to Muszynski et al. [42]

EMC	Compression 30%	Compression 90%	Tension
E_{MS} [GPa]	37.8	8.88	38.1
m_1 [GPa^{-1}]	0.48	5.78	0.504
m_2 [GPa^{-1}]	0.04	0.04	0.005

By extrapolating this equation to a period of 50 years, the constant slope governs the deformation, which leads to larger creep coefficients compared to the measured ones. However, using a power function to fit the results given in Muszynski et al. [42] by the method of least squares (see Fig. 6.6(a)), the increases of the creep strains for moisture accumulations larger than 65% are not as large as in the equation given by Muszynski et al. [42].

$$\varepsilon_{ms} = d \cdot t^b \cdot \varepsilon_{ms}(\nu = 65\%) \text{ for } \sum u > 20\% \qquad (6.2)$$

where $\varepsilon_{ms}(\nu = 65\%)$ Mechano-sorptive creep at a accumulated moisture content of 65%
d = 0.5756
b = 0.1252

If the increase of the mechano-sorptive creep strain in the studies according to Muszynski et al. [42] i. e. the fitted equation based on these test results and the models according to Toratti [51] are compared (see Fig. 6.7), the correlation between the fitted equation Eq. (6.2) of test results according to Muszynski et al. [42] and model B is better than between the fitted equation Eq. (6.2) of the test results and model D. Based on these tests, model B should be sufficient to model the mechano-sorptive creep.

If the mechano-sorptive creep governs the creep deformations and no mechano-sorptive creep limit has been reached, the cross section dimensions influence the creep deformation, since the average accumulated moisture content decreases with increasing thickness due to the moisture penetration process. When comparing the creep coefficients of the round wood elements determined by measurements to each other, no decreasing creep coefficient with increasing cross section diameter can be found. Instead, the creep coefficient increases with an increasing diameter of the cross section (see Fig. 6.8). This increase can among others be explained by an increased stiffness of the beam. Since the measurements are performed within the structure and interactions between the single structural elements exist, the stiffer

6.1 General

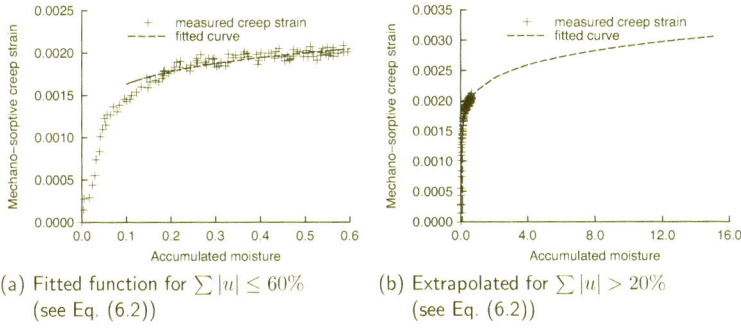

(a) Fitted function for $\sum |u| \leq 60\%$ (see Eq. (6.2))

(b) Extrapolated for $\sum |u| > 20\%$ (see Eq. (6.2))

Figure 6.6: Modeling the mechano-sorptive creep strain based on the values given in Muszynski et al. [42]

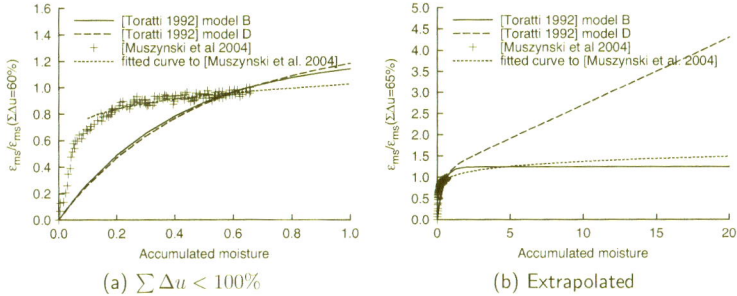

(a) $\sum \Delta u < 100\%$

(b) Extrapolated

Figure 6.7: Comparison of the creep strain related to the mechano-sorptive creep strain at $\sum \Delta u = 65\%$

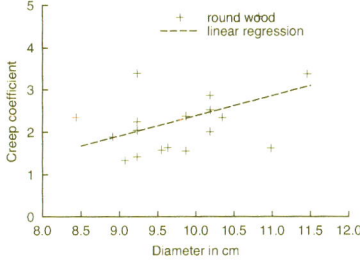

Figure 6.8: Influence of the cross section diameter on the creep coefficient

elements attract forces. Therefore, it is very likely, that the loading of the stiffer elements is larger than that of the softer elements, which results in higher creep coefficients. The softer elements can reduce their loading, so their creep coefficients are smaller.

To summarize, tests results according to Mohager [39] show a strong dependence on the accumulated moisture content. This accumulation of the tests is comparable to the evaluated accumulated moisture content for the elements in the region of Tübingen. However, a dependence of the creep coefficient on the cross section cannot be verified by measurements. Besides that, the studies according to Muszynski et al. [42] tend to validate the mechano-sorptive creep strain of model B. Therefore it is assumed, that for an usual stress level, some kind of mechano-sorptive creep limit is reached. So the modeling of the mechano-sorptive creep based on the model B should be sufficient in these cases.

6.2 Modification of model B according to Toratti [51]

In order to describe the time-dependent behavior of timber, model B according to Toratti [51] is modified (see Fig. 6.9), since this model shows the smallest differences to the measured results.

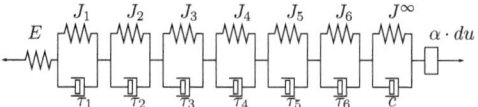

Figure 6.9: Model B according to Toratti [51]

It is assumed that the differences are caused by the normal creep due to the reasons given in the previous section. If the parameters are fitted to the results of the measurements by the method of least squares, the parameters given in Tab. 6.2 are obtained.

Table 6.2: Parameters

Original values element	1	2	3	4	5	6
τ_i	0.01	0.1	1	10	100	5000
J_i	0.0686	-0.0056	0.0716	0.0404	0.2073	0.5503
Modified values element	1	2	3	4	5	6
τ_i	0.01	0.1	1	10	193.23	11078.51
J_i	0.0686	-0.0056	0.0716	0.0409	0.2201	1.8052

Since only elements 4 to 6 have to be modified in order to get a sufficient correlation between measurements and evaluation, only the prediction for a period of more than 10 years is influenced by the modification (see Fig. 6.10 and Fig. 6.11). Therefore all performed verifications in the period of less than 10 years are still valid.

6.3 Comparison to measurements in the region of Breisgau-Hochschwarzwald

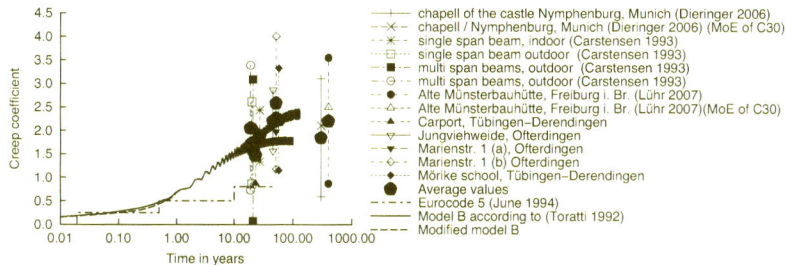

Figure 6.10: Comparison of the modified model B according to Toratti [51] to the measured creep coefficients

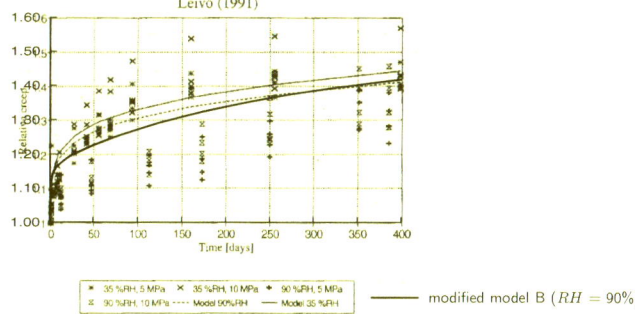

Figure 6.11: Comparison of model B according to Toratti [51] and modified model B, respectively, with tests results (taken from Toratti [51] and added creep coefficient based on modified model B)

6.3 Comparison to measurements in the region of Breisgau-Hochschwarzwald

Gutenkunst [20] performed measurements similar to those mentioned in Chap. 5 around Löffingen in the region of Breisgau-Hochschwarzwald (see Fig. 6.12). In contrast to the measurements in the region of Tübingen the snow in this region cannot be neglected, since the average duration of snow is 84 days per year and the average snow height within this period is 15cm, which should be equivalent to a snow load of $0.4 kN/m^2$.

For the determination of the creep coefficient, Gutenkunst [20] measured the current deformation of three different buildings located in the region around of Löffingen, Breisgau-Hochschwarzwald in the villages of Behla, Reiselfingen and Unadingen (see Fig. 6.13).

The years of construction of these buildings are between 1968 and 1977. All buildings have in common, that they are buildings without any heating. So they fulfill the requirements for

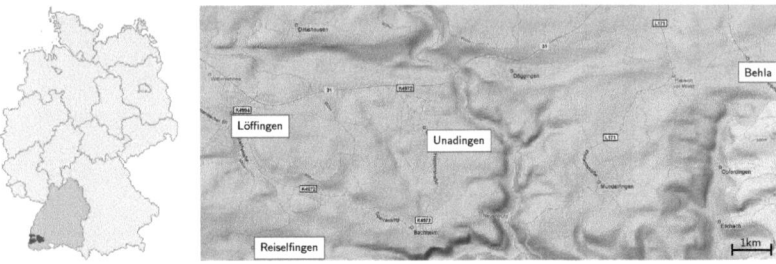

Figure 6.12: Location of Löffingen (see wikipedia.de [54] and maps.google.de [37])

(a) Behla, 725m above sea level, 47°53'51.26"N, 8°30'46.88"E

(b) Reiselfingen, 765m above sea level, 47°51'5.37"N, 8°21'5.61"E

(c) Unadingen, 774m above sea level, 47°52'51.67"N, 8°24'19.63"E

Figure 6.13: Buildings, where the measurements where performed by Gutenkunst [20]

the measurements (see Sec. 5.2.2). The dimensions and structural systems of the studied buildings are given in Tab. 6.3.

The surrounding conditions and the course of the snow load are modeled by the daily values given by Deutscher Wetterdienst [6] of a weather station in Lenzkirch-Ruhbühl at a distance of about 10km from the measured buildings. These data provide the relative humidity, the

6.3 Comparison to measurements in the region of Breisgau-Hochschwarzwald

Table 6.3: Measured elements

	Behla	Reiselfingen	Unadingen
structural system	single span	two span	single span
span	3.93m	2.56m & 2.67m	4.05m
cross section dimensions	9.5×13.5cm	7.5×13.5cm	7.5×15cm
year of construction	1968	1971	1977
numbers of accepted elements	4	4	5

snow height and the temperature (see Fig. 6.14 and Fig. 6.15). Unfortunately, no values

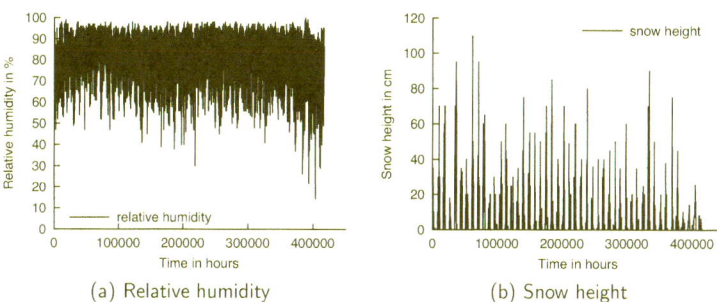

(a) Relative humidity   (b) Snow height

Figure 6.14: Surrounding conditions in Löffingen, region of Breisgau-Hochschwarzwald

are given for June 1996, for the period between September 01, 2004 and March 31, 2007 and from August 1, 2007 until now. In order to fill these gaps, the data of the weather station of Feldberg at a distance of 20km are used. For the relative humidity, it is assumed, that the values of the weather station Feldberg can be used directly, whereas the snow load differs, since the elevation of this weather station is much higher than the weather station in Lenzkirch-Ruhbühl. For this reason, the snow load of Feldberg is taken and multiplied by a factor, which gives the ratio between the accumulated snow height in the years 2000 to 2004 in Lenzkirch-Ruhbühl and in Feldberg.

$$\sum_{2000}^{2004} h_{snow,\text{Lenzkirch-Ruhbühl}} \cdot \Delta t = k_{red} \cdot \sum_{2000}^{2004} h_{snow,\text{Feldberg}} \cdot \Delta t \tag{6.3}$$

In order to determine the snow load, a density of snow of 350kg/m^3 is assumed. This density represents wet snow (see [45]). In Fig. 6.15 the course of the temperature in Lenzkirchen-Ruhbühl of January 1, 1961 is shown. As can be seen, there are cold days, but if the course of the temperature is approximated by Fourier series where only amplitudes with $\Delta T_i \geq$ 1°C are considered, the minimum temperature is 2.3°C. This indicates, that no long periods of frost exist in this region. Therefore the snow settles and gets wet, resulting in a higher density.

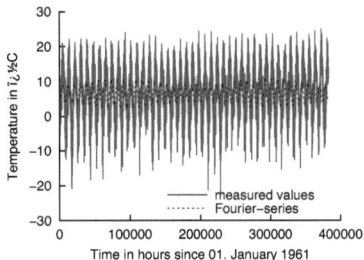

Figure 6.15: Temperature in Lenzkirchen

For the evaluation of the temporal deformation based on the modified model B according to Toratti [51] the daily relative humidity as well as the daily snow height are considered (see Fig. 6.14), assuming, that the date of erection is the 1st of June in the year of construction. In Fig. 6.16 the evaluated creep coefficients and measured average creep coefficients with their standard deviations of the three buildings in the region of Breisgau-Hochschwarzwald are shown. The peaks in these figures are caused by the deformation due to snow since the

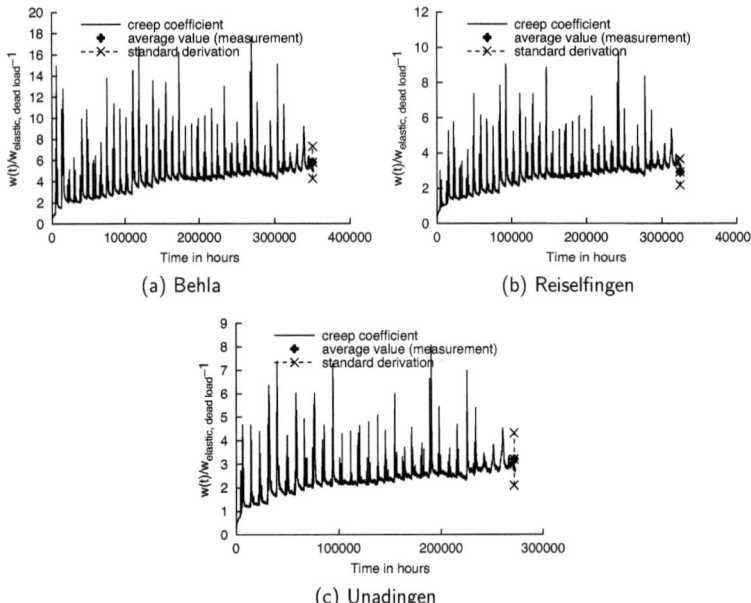

(a) Behla

(b) Reiselfingen

(c) Unadingen

Figure 6.16: Comparison between evaluated and measured creep coefficients, which are related to the elastic deformation due to permanent dead load

6.4 Conclusions

creep coefficient is determined by

$$\varphi = \frac{w_{\text{tot}}}{w_{\text{elastic, dead load}}} - 1 \qquad (6.4)$$

where w_{tot} total deflection
                       $= w_{\text{elastic, dead load}} + w_{\text{elastic, snow load}} + w_{\text{creep, dead load}} + w_{\text{creep, snow load}}$
       $w_{\text{elastic, dead load}}$ elastic deformation for the dead load of the roof

As shown in Fig. 6.16, the evaluated creep coefficients fit the measured ones. Therefore the modified model B according to Toratti [51] seems applicable in normal surrounding conditions in the region of South-West Germany.

6.4 Conclusions

As the comparison of the measured values in the region of Tübingen and in the region of Breisgau-Hochschwarzwald and the evaluated deformations with the modified model B according to Toratti [51] shows, the model can be used for the prediction of the long-term deformation in the South-West of Germany. Since the model fits in the region of Tübingen as well as in the region of Breisgau-Hochschwarzwald, although the accumulated relative humidity between both regions differs (see Fig. 6.17), it can be concluded, that the mechano-sorptive creep seems to reach a creep limit in normal surrounding conditions.

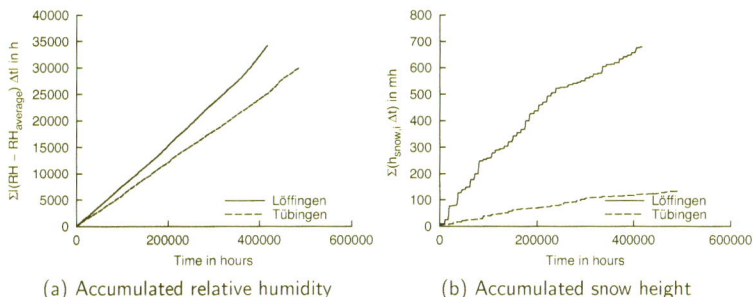

Figure 6.17: Accumulated relative humidity and accumulated snow height in the region of Tübingen and Breisgau-Hochschwarzwald

However, there is still this discrepancy, that the used model B does not correspond to the test results according to Mohager [39] (see Fig. 6.2), where faster and heavy moisture variations take place compared to real moisture variations in South-West Germany. For this reason, the modified model B seems to be applicable only for "real" moisture variations as can be found in South-West Germany but not for all possible moisture variations which may be typical in other regions. Therefore, no general use of this model can be recommended.

One maybe possible approach to solve the discrepancy between the tests by Mohager [39] and the model B is given in Appendix B. Within this proposal, the creep behavior is evaluated by a bionic approach based on the explanatory model by Boyd [4]. Within this proposal the mechano-sorptive creep is caused by moisture induced eigenstresses. Therefore not only the absolute moisture variation but also the rate of the moisture variation influence the mechano-sorptive creep strain. Due to the dependence on the moisture variation as well as on the rate of the moisture variation the differences between the tests by Mohager [39] and the measurements can be explained in principle, since in the tests by Mohager [39] the timber is exposed to fast and heavy moisture variations. This leads to higher eigenstresses in the microfibril and therefore to larger creep strains than in the measured elements. However there is still the need to improve and validate the model to additional configurations in order to generalize the application of the model.

7 Simplified equations for the determination of creep deformations

7.1 General

In order to determine the time-dependent deformations, it has been shown, that the model B according to Toratti [51] and the modified model B, respectively, are applicable to evaluate the creep deformations. However, the use of these models seems to be complex. In the following, simplified equations of the creep coefficients of timber in constant as well as in variable climate are determined, which consider the most important influences. By these simplified equations, it should be much easier to determine the creep coefficients and to consider the various influences on the creep coefficient.

For the determination of the time-dependent deflection the different effects will be classified into two groups:

- global creep effects
- local creep effects

The group "global creep effects" considers all the effects that are more or less equal in every point of the cross section. The creep coefficient, belonging to this group, will describe the normal creep, the mechano-sorptive creep due to drying and the mechano-sorptive creep due to minimum changing of the moisture content in the cross section. Since it is assumed, that the initial moisture content is constant over the cross section, the difference in the moisture content between initial and equilibrium moisture content is the same in every point in the cross section. Therefore the mechano-sorptive creep due to drying to the equilibrium moisture content in the cross section is more or less equal in each point.

In order to determine the time-dependent deflection, all the global creep effects caused by drying are summed up into an effective creep coefficient by the following equation

$$\varphi = \varphi_0 \cdot DSF \tag{7.1}$$

where φ_0 basic creep coefficient for constant moisture
DSF factor considering the influence of drying

A rule, comparable to the factor DSF, has already been introduced by the standards DIN 1052 [9] or Eurocode 5 [14]. Within these standards, the creep coefficient has to be increased by the addition of 1.0, if the moisture content of the cross section is around or above the fibre saturation point and the cross section can dry in the structure. So the DSF-factor according to the mentioned standards would be evaluated according to following equation.

$$DSF_{\text{DIN 1052 [9]/Eurocode 5 [14]}} = 1 + \frac{1}{k_{def}} \tag{7.2}$$

where $DSF_{\text{DIN 1052 [9]/Eurocode 5 [14]}}$ parameter DSF according to the standards DIN 1052 [9] and Eurocode 5 [14], respectively
k_{def} creep coefficient according to the standards

In addition to the normal creep and the influence of drying, changing moisture can lead to a global change of the moisture content within a moisture cycle. If the cross section is so small that the moisture content in the middle of the cross section is influenced by the mechano-sorptive creep, this mechano-sorptive creep has to be considered in the global creep coefficients.

In order to get values for the mechano-sorptive creep which appears at least in every point of the cross section, the minimum variation of the changing moisture content in a cross section has to be determined. This minimum moisture variation occurs at least in each point in the cross section.

Since most of the models introduce creep limits for the pure creep as well as for the mechano-sorptive creep, a superposition just by adding the influences is hardly possible. Therefore the interaction between drying and annual cycling moisture variation will be considered by an effective moisture variation due to drying. In order to determine the effective moisture variation, the effects of drying are transformed into effects caused by an annual cycling of the moisture content. The effective moisture variation considers the real annual moisture variation and the fictitious moisture cycle due to drying. With this effective moisture variation, the creep coefficient can be determined by

$$\varphi = GMS \cdot \varphi_0 \qquad (7.3)$$

where φ_0 basic creep coefficient for constant moisture
GMS factor considering the influence of **g**lobal **m**echano-**s**orptive creep

In the second group – the "local creep effects" – all effects are summed up, which are not constantly distributed over the whole cross section, as for example the influence of the moisture variation resulting in mechano-sorptive creep. However, these influences depend on the local moisture variation. Therefore the thickness of the cross section should not influence this local part of the moisture distribution, as long as there is no moisture variation in the center of the cross section.

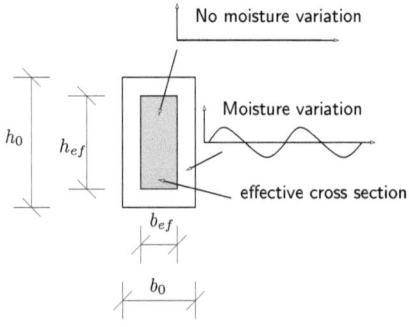

Figure 7.1: Principal consideration of changing moisture

In order to consider this group of creep deformation in a practical design procedure, the creep deformations of this group are considered by effective cross section dimensions (see Fig. 7.1). So it is thinkable, that for a certain yearly change of the relative humidity the mechano-sorptive creep in the outer layers leads to such high local creep deformations that these outer layers hardly participate in the load transfer. The thickness of this non-participating layer has to be determined.

For practical use, the real cross section may be reduced to an effective cross section, just by subtracting this ineffective layer, comparable to the design for fire, where the charring layers are neglected in the design. The deformation can be determined by using this reduced cross section in combination with the global creep coefficient, so all effects are considered.

The complete creep deformation of a cross section consists of the creep coefficient assumed as constant over the cross section and the reduced height for considering non-constant distributed creep deformations (see Fig. 7.2).

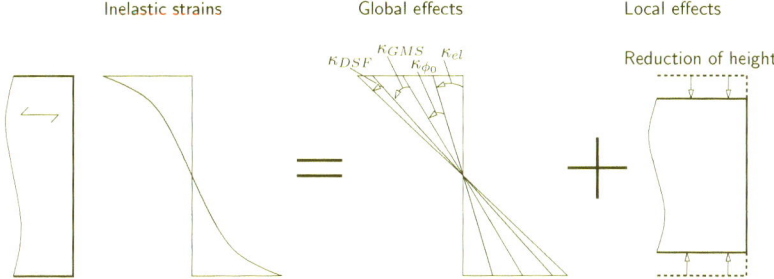

Figure 7.2: Concept of the consideration of time-dependent effects

7.2 Global creep effects based on model B according to Toratti [51] and the modified model B

7.2.1 Creep coefficients for constant environmental conditions

For the determination of the creep coefficients based on model B according to Toratti [51] in a constant climate the original equation (comp. Eq. (3.11)) can be used. This is based on the fact, that the rheological model has been adapted to this equation (see Fig. 7.3). The parameters for the modified model B are evaluated by fitting the rheological model to the measured results. Based on this model, a power function is fitted to the evaluated normal creep strain.

Transforming Eq. (3.11) or fitting, respectively, a power function to the modified model B the following equation for the creep coefficient in constant humidity may be determined by

$$\varphi_0 = \frac{E(u)}{E(u_{ref})} \cdot \left(\frac{t}{t_d}\right)^k \tag{7.4}$$

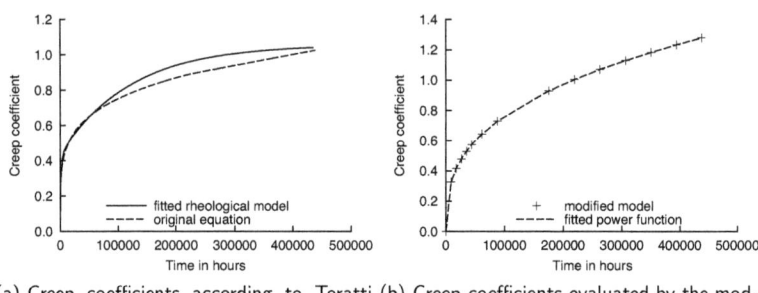

(a) Creep coefficients according to Toratti [51]

(b) Creep coefficients evaluated by the modified model B

Figure 7.3: Evaluated creep coefficients by means of the model B according to Toratti [51] and the modified model B

where t current time
t_d, k parameters according to Tab. 7.1
u_{ref} =0.2
$E(u)$ Modulus of Elasticity at moisture content u

Table 7.1: Parameters for the creep coefficients in constant surrounding conditions evaluated by the model B according to Toratti [51] and by the modified model B

	model B	modified model B
t_d	29500 days	8994 days
k	0.21	0.35

If the relation of the Modulus of Elasticity is introduced according to Eq. (3.17), the creep coefficient can be determined by (see Fig. 7.4)

$$\varphi_0 = \frac{1 - 1.06 \cdot u}{0.788} \cdot \left(\frac{t}{t_d}\right)^k \tag{7.5}$$

Within this determination of the creep coefficient, the stress level does not influence the creep coefficient in constant humidity, whereas an increasing moisture content leads to a decreasing creep coefficient.

7.2.2 Consideration of the changing moisture while reaching the equilibrium moisture content after 50 years

In order to consider the effects of the drying process on the time-dependent behavior a **drying shift factor** DSF is introduced. This factor considers the effects of mechano-sorptive creep, of different shrinkage and swelling and of the changes of the Modulus of Elasticity.

7.2 Global creep effects based on model B according to Toratti [51] and the modified model B

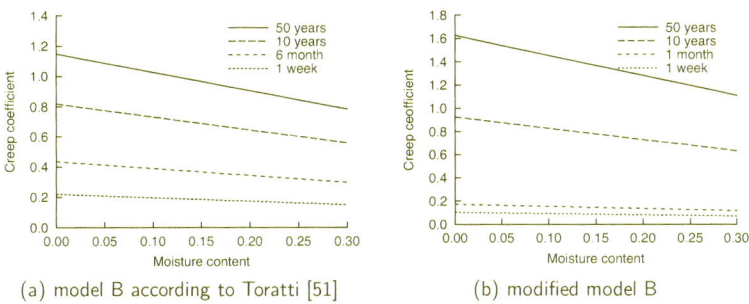

(a) model B according to Toratti [51]   (b) modified model B

Figure 7.4: Creep coefficients based on the model B according to Toratti [51] and the modified model B (see Eq. (7.5))

This factor DSF is determined by comparing the results of a drying process using the developed tool *kriHo* to the results of Eq. (7.5). The difference between these two results is considered within a drying shift factor (see Fig. 7.5). The drying shift factor depends only on the difference between the initial and the equilibrium moisture content, since the strain in the model according to Toratti [51] and the modified model B is linearly dependent on the stress.

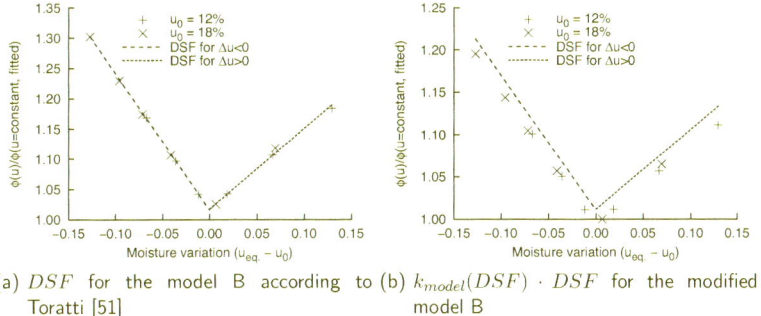

(a) DSF for the model B according to Toratti [51]   (b) $k_{model}(DSF) \cdot DSF$ for the modified model B

Figure 7.5: Drying shift factor for the model according to Toratti [51] and the modified model B, respectively, according to Eq. (7.6) (for $k_{model}(DSF)$ see Eq. (7.10))

A linear function for the determination of the drying shift factor seems to be sufficient (see Fig. 7.5). For the rheological model B according to Toratti [51] and the modified model B, respectively, the factor DSF can be determined to be

$$DSF(50 \ years) = \frac{\varphi(u + \Delta u_{drying})}{\varphi(u = constant)} = m \cdot |\Delta u| + c \qquad (7.6)$$

where m,c coefficients according to Tab. 7.2
Δu difference between equilibrium and initial moisture content
$u_{\text{equilibrium}} - u_0$

Table 7.2: Coefficients for the drying shift factor

	m	c
$\Delta u > 0$	1.334029791	1.017635116
$\Delta u < 0$	2.252280126	1.01566572

The creep coefficient according to Toratti [51] and the modified model B, respectively, under consideration of the drying process may be determined by

$$\varphi = k_{model}\left(m \cdot |\Delta u| + c\right) \cdot \left(m \cdot |\Delta u| + c\right) \cdot \frac{1 - 1.06 \cdot u_0}{0.788} \cdot \left(\frac{t}{t_d}\right)^k \tag{7.7}$$

where t time of 50 years
t_d, k parameters of the model according to Tab. 7.1
m,c coefficients according to Tab. 7.2
Δu difference between equilibrium and initial moisture content
$u_{\text{equilibrium}} - u_0$
$k_{model}(DSF)$ model parameter to consider which of the models (model B according to Toratti [51] or modified model B) is used (see Eq. (7.10)

The parameter k_{model} considers, whether the model B according to Toratti [51] or the modified model B according to Chap. 6 is used. This parameter can be determined analytically, since the only difference between those two models is the normal creep, whereas the mechano-sorptive creep is equal in both models. Since all creep strains are related to the normal creep strain and the total creep strain is the sum of normal creep strain and mechano-sorptive creep strain,

$$C_{infl} \cdot \varepsilon_{\text{normal creep}} = \varepsilon_{\text{normal creep}} + \varepsilon_{\text{mechano-sorptive creep}} \tag{7.8}$$
$$\varepsilon_{\text{mechano-sorptive creep}} = (C_{infl} - 1) \cdot \varepsilon_{\text{normal creep, model B}}$$
$$= (k_{model} \cdot C_{infl} - 1) \cdot \varepsilon_{\text{normal creep, modified model B}} \tag{7.9}$$

where C_{infl} DSF or GMS
the model parameter k_{model} can be evaluated by

$$k_{model}(C_{infl}) = \left(\begin{array}{ll} \frac{1}{t^{-k_m+k} \cdot t_{0m}^{k_m} \cdot t_0^{-k}} + \frac{(-t_m)^{-k_m+k} \cdot t_{0m}^{k_m} \cdot t_0^{-k}+1}{C_{infl}} & \text{[51] model B} \\ & \text{modified model B} \end{array}\right) \tag{7.10}$$

7.2 Global creep effects based on model B according to Toratti [51] and the modified model B

where
- t period of time ($=50$ years)
- t_0 parameter of the model
- k parameter of the model
- C_{infl} DSF or GMS
- subscript m parameters of the modified model B
- without subscript   parameters of the model B according to Toratti [51]

Since the parameters of the models k, k_m, t_0 and t_{0m} (see Tab. 7.1) as well as the point in time ($t = 50$ years) after loading are defined, the determination of the model parameter k_{model} can be simplified to

$$k_{model}(C_{infl}) = \begin{pmatrix} 1 & \text{for [51] model B} \\ 0.7057 + \frac{0.2942}{C_{infl}} & \text{for modified model B} \end{pmatrix} \quad (7.11)$$

7.2.3 Influence of the moisture content, changing in one-year sinusoidal cycles

7.2.3.1 Determination of the minimum moisture variation within the cross section

Changing surrounding conditions influence the moisture content within the cross section. If there is a large change of the relative humidity especially in slender cross sections, the moisture content even in the middle of the cross section is influenced by the changing relative humidity. Therefore mechano-sorptive creep can even arise not only in the outer layers but also within the cross section. In order to fit the concept of local and global creep, the minimum mechano-sorptive creep of the cross section has to be determined. Since the mechano-sorptive creep depends on the changing moisture content, the variation of the moisture content has to be determined.

The idea is to determine the minimum variation of the moisture content within a moisture cycle in the cross section. With this minimum variation of the moisture content, the minimum mechano-sorptive creep can be determined. That means, that at least this minimum mechano-sorptive creep arises in each point of the cross section (see Fig. 7.6). Therefore this

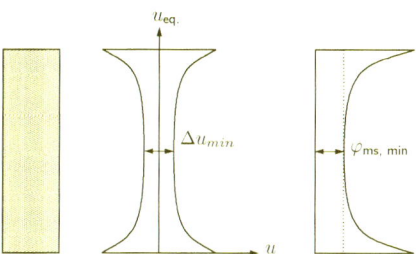

Figure 7.6: Minimum moisture content in the cross section and idealized distribution of the creep coefficient due to mechano-sorptive creep

minimum mechano-sorptive creep is independent of the dimensions but only dependent on the moisture variation. The basic creep coefficient can be increased in order to consider this additional creep. For determining the minimal moisture variation a case study is performed, based on following assumptions:

- only a one dimensional moisture variation (see Fig. 7.7)

- no memory effects, so the minimum or maximum moisture content within the first 5 years is the same as within the lifetime of the cross section.

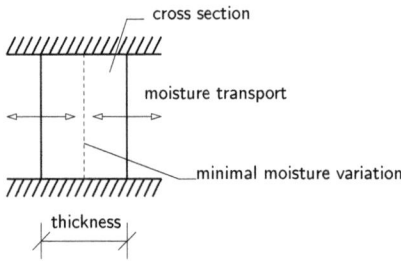

Figure 7.7: Assumed moisture transportation

If the changing moisture is related to the changing humidity (see Fig. 7.8), it becomes obvious, that the determination of the moisture variation can be split into two functions; one function describing the influence of the thickness of the cross section and the other function determining the influence of the value of the moisture variation. In this function only the change of the humidity influences the change of the moisture content; the influence of the moisture level seems to be negligible.

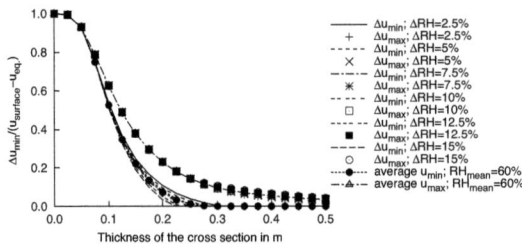

Figure 7.8: Influence of the thickness on the minimal moisture variation (=moisture content in the center of the cross section)

The moisture variation depends on the thickness and the air moisture variation. It can be determined in dependence on the direction of the moisture change, since the permeability according to the model of Toratti [51] is dependent on the moisture content.

7.2 Global creep effects based on model B according to Toratti [51] and the modified model B

- decrease of the moisture content

$$\frac{\Delta u_{min}(b)}{\Delta u_{air}} = \begin{cases} 20.68 \cdot b^2 - 11.15 \cdot b + 1.48 & \text{for } b < 0.30m \\ \leq 1.0 \\ \geq 0 \\ 0 & \text{for } b > 0.30m \end{cases} \quad \text{for } \Delta u_{air} < 0 \quad (7.12)$$

where b thickness of the cross section in m

- increase of the moisture content

$$\frac{\Delta u_{max}(b)}{\Delta u_{air}} = \begin{cases} 14.44 \cdot b^2 - 8.32 \cdot b + 1.31 & \text{for } b < 0.30m \\ -0.27 \cdot b + 0.17 & \text{for } b > 0.30m \\ \leq 1.0 \\ \geq 0 \end{cases} \quad \text{for } \Delta u_{air} > 0 \quad (7.13)$$

where b thickness of the cross section in m

By this equation, the minimal moisture content of a cross section with an one dimensional moisture transport can be determined. Since the relative humidity is often known, the differences of the moisture content can be determined as follows

1. determination of the equilibrium moisture content according to Eq. (3.21)
   $\rightarrow u_{eq.}$

2. determination of the maximum and minimum moisture content according to Eq. (3.21) with the input value $RH \pm \Delta RH$
   $\rightarrow u_{min,air}$ and $u_{max,air}$, respectively

3. determination of the equilibrium moisture content in the changing surrounding climate (see Fig. 7.9)
   $\rightarrow \Delta u_{min,air} = u_{min,air} - u_{eq.}$ and $\Delta u_{max,air} = u_{max,air} - u_{eq.}$, respectively

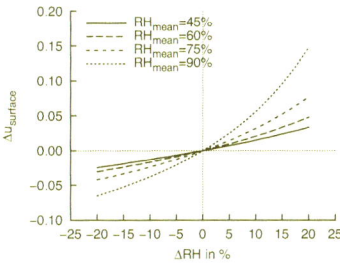

Figure 7.9: Relation between change of relative humidity and equilibrium moisture content

4. determination of the minimum changing moisture content of the cross section by Eq. (7.12) and Eq. (7.13) during the one-year sinusoidal cycle of the relative humidity
   $\rightarrow \Delta u_{min}(b)$ resp. $\Delta u_{max}(b)$

5. determination of the total moisture change by
   $\Delta u = |\Delta u_{min}(b)| + \Delta u_{max}(b)$

7.2.3.2 Determination of creep coefficients considering the global mechano-sorptive creep

In order to determine the increase of the creep deformation due to a changing moisture, the coefficient GMS is introduced, which describes the ratio between the creep coefficient in the changing surrounding conditions and the creep coefficient in the constant surrounding conditions.

$$GMS = \frac{\varphi(u(t))}{\varphi(u = \text{const.})} \tag{7.14}$$

For the determination of the creep coefficient, very slender cross sections with a thickness of 1cm are evaluated. As seen in Fig. 7.8 only little differences between the minimum moisture content and the air moisture content arise.

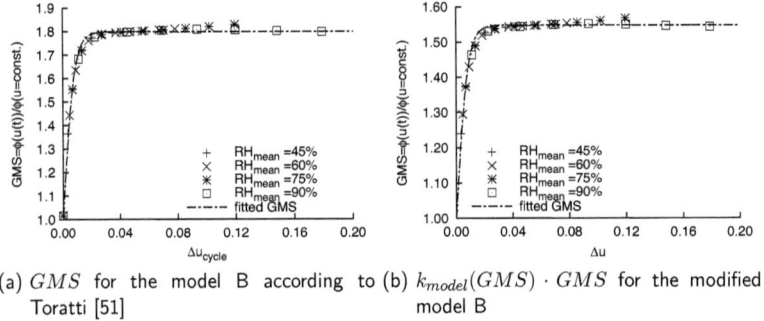

(a) GMS for the model B according to Toratti [51]

(b) $k_{model}(GMS) \cdot GMS$ for the modified model B

Figure 7.10: Dependence of the coefficient GMS on the changing moisture content Δu for the model B according to Toratti [51] and the modified model B, respectively

By fitting an equation to the results by the method of least squares, this coefficient GMS can be determined to be

$$GMS = 0.783 \cdot tanh\,(124.105 \cdot \Delta u) + 1.015 \tag{7.15}$$

So the creep coefficient can be determined by

$$\varphi = \varphi_0 \cdot GMS(\Delta u_{\text{cycle}}) \cdot k_{model}(GSM) \tag{7.16}$$

where	φ_0	basic creep coefficient for constant moisture according to Eq. (7.5)
	GMS	factor considering the influence of global mechano-sorptive creep according to Eq. (7.15)
	Δu_{cycle}	minimum moisture variation within a moisture cycle
	$k_{model}(GMS)$	model parameter to consider which of the models (model B according to Toratti [51] or modified model B) is used (see Eq. (7.10))

7.2.4 Interaction between drying and changing moisture

Since the mechano-sorptive creep reaches a limit in the model according to Toratti [51] (see Fig. 7.10), the influence of the drying and the mechano-sorptive creep cannot be summed up. Therefore the idea is, to transform the drying process into an effective moisture variation. The increase of the creep coefficients due to drying and changes of the relative humidity can be described by Eq. (7.15).

This means, that the effective moisture variation can be determined by the equality of the coefficient DSF and the GMS

$$k_{model}(DSF) \cdot DSF(\Delta u) \equiv k_{model}(GMS) \cdot GMS(\Delta u_{eff}) \qquad (7.17)$$

This equation can be simplified to

$$DSF(\Delta u) \equiv GMS(\Delta u_{eff}) \qquad (7.18)$$

since, if Eq. (7.18) is fulfilled, the parameters $k_{model}(DSF)$ and $k_{model}(GMS)$ are equal. Therefore the determination of the effective moisture variation Δu is valid for the model B according to Toratti [51] as well as for the modified model B.

Using Eq. (7.18) following effective Δu can be determined

- for $u_{eq.} > u_0$

$$\Delta u_{eff} = \frac{1}{124.05} \cdot ln\left(\frac{0.785 + 2.2522 \cdot \Delta u_{drying}}{0.785 - 2.2522 \cdot \Delta u_{drying}}\right) \qquad (7.19)$$

- for $u_{eq.} < u_0$

$$\Delta u_{eff} = \frac{1}{124.05} \cdot ln\left(\frac{0.785 + 1.3340 \cdot \Delta u_{drying}}{0.785 - 1.3340 \cdot \Delta u_{drying}}\right) \qquad (7.20)$$

where	$u_{eq.}$	equilibrium moisture content
	u_0	initial moisture content
	Δu	$u_0 - u_{eq.}$

From this, the effective change in moisture due to drying can be added to the moisture variation caused by the annual cycle of the relative humidity in order to determine the effective creep coefficient.

7.3 Local creep effects

7.3.1 General

A change in moisture in combination with stress leads to an increase of the time-dependent deformation. Since the moisture variation is not constant over the cross section (see Fig. 7.11), the effects are dependent on the local changing of the moisture content.

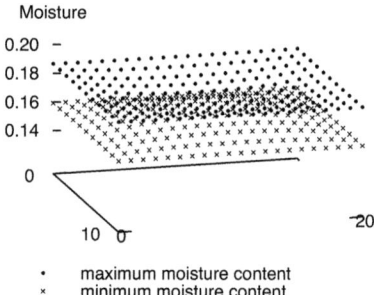

- maximum moisture content
- × minimum moisture content

Figure 7.11: Maximum and minimum moisture content, respectively, of a 100mm × 200mm timber beam, subjected to an annual cycle of the relative humidity with an amplitude of 10%

In order to determine the local increase of the creep deformation due to the local moisture variation, a reduction of the height will be introduced. This follows the assumption that the deformation f, taking the moisture variation into account, is equal to the deformation f of the reduced cross section but with a creep coefficient, only considering the normal creep and the minimum mechano-sorptive creep in the whole cross section, e.g. for a single span girder:

$$f = \frac{5 \cdot q \cdot l^4}{384 \cdot E \cdot J} \cdot (1 + \varphi_{\text{changing moisture}}) \equiv \frac{5 \cdot q \cdot l^4}{384 \cdot E \cdot J_{ef}} \cdot (1 + \varphi_{\text{normal + minimum ms creep}}) \quad (7.21)$$

Transforming this equation, the following equations for the reduction of the height resp. width can be determined by

$$J_{eff} = J \cdot \frac{1 + \varphi_{\text{normal + minimum ms creep}}}{1 + \varphi_{\text{changing moisture}}} = \frac{(h - 2 \cdot \Delta s)^3 \cdot (b - 2 \cdot \Delta s)}{12} \quad (7.22)$$

Unfortunately, the equation cannot be solved analytically in a reasonable way, so the solution is done numerically.

7.3.2 Determination of effective dimensions for the model according to Toratti [51]

In order to consider the expected increase of the creep deformation due to the increased moisture variations in the outer layers, this concept is applied to the results of a 100mm×200mm timber beam.

However, in the end, the increased moisture variation at the outer layer does not seem to influence the time-dependent deflection according to the model of Toratti [51] and in the modified model B, because the creep coefficients, evaluated by *kriHo* in consideration of the increased moisture variation, and the creep coefficients according to Eq. (7.16) do not differ essentially (see Fig. 7.12). The reason is, that in the model according to Toratti [51]

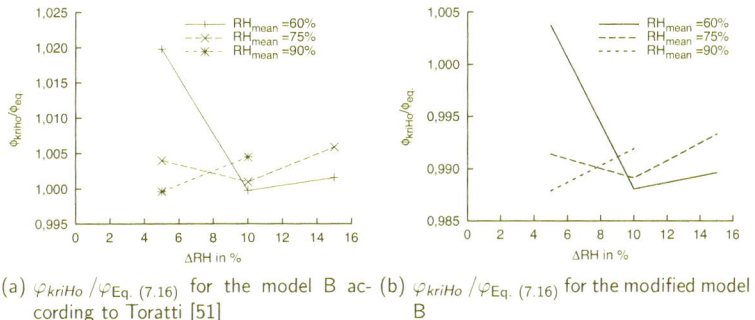

(a) $\varphi_{kriHo}/\varphi_{Eq.\ (7.16)}$ for the model B according to Toratti [51]

(b) $\varphi_{kriHo}/\varphi_{Eq.\ (7.16)}$ for the modified model B

Figure 7.12: Comparison of the creep coefficients in changing humidity

φ_{kriHo} = creep coefficients considering the increased moisture variations in the outer layers

$\varphi_{eq.}$ = creep coefficients considering "only" the global mechano-sorptive creep

the mechano-sorptive creep reaches a limit. Therefore the differences between the creep coefficients considering the increased moisture variations in the outer layers and the creep coefficients considering "only" the global mechano-sorptive creep is less than 2% for the model B and less than 0.5% for the modified model B (see Fig. 7.12).

In Fig. 7.13 the required annual amplitude of the relative humidity is shown, for which the GMS-value according to Eq. (7.15) is 95% of the GMS-value for the maximum moisture variation. In this case the differences in the mechano-sorptive creep in the outer layers $(=GMS_{max}) \cdot \varphi)$ is comparable to the mechano-sorptive creep in the inner layers $(=GMS(\Delta RH_{min}) \cdot \varphi)$. As shown in Fig. 7.13 only little annual variations of the relative humidity are necessary in order to reach the creep limit in the model according to Toratti [51] and the modified model. Therefore, even for larger dimensions, the mechano-sorptive creep is comparable over the whole cross section, if the cross section is subjected to service class 2 conditions.

Since the mechano-sorptive creep reaches its creep limit, no differences in the mechano-sorptive creep coefficient within the cross section can be found. Therefore – in contrast to the other models (see Appendix C) – no effective height is necessary for the creep coefficients

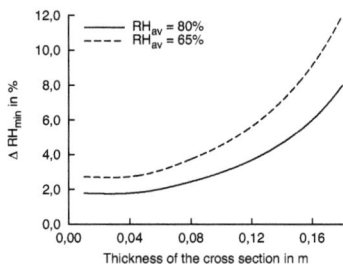

Figure 7.13: Minimum amplitude of the annual cycle of the relative humidity ΔRH_{min} for $GMS(\Delta RH_{min}) = 0.95 \cdot GMS_{max}$)

based on the model B according to Toratti [51] and the modified model B.

8 Influence of the load history

8.1 General

Since creeping is a time-dependent process, the duration of the load is of importance. In order to determine this influence, the models are subjected to a load history, given in Fig. 8.1. These load histories have in common that a live load acts temporary. Therefore the resulting

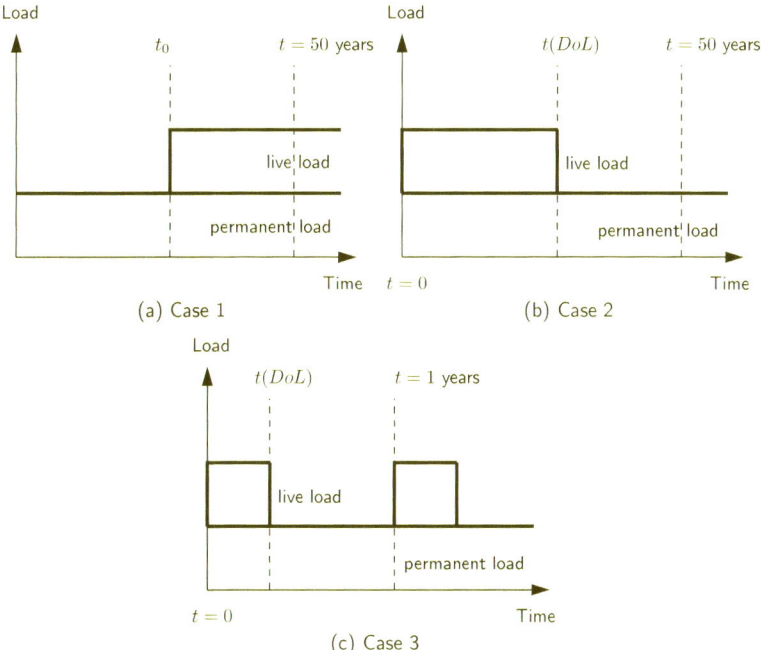

Figure 8.1: Different scenarios concerning the load history

creep coefficient cannot be determined directly, since in terms of unloading a recovery of the deflection may take place. Therefore the question is, whether a live load causes creep deformations or not.

In view of a practical use, in principle two different approaches for the consideration of the non-permanent loads exist:

- creep coefficient for each duration of load: In this concept, every duration of load has its own creep coefficient (see a former version of Eurocode 5 [14]). Unfortunately, this procedure leads to extensive calculations since in the semi-probabilistic design, the different loads have to be combined with their combination coefficient ψ_i and additionally the different creep coefficients have to be considered for the determination of the deflections. Besides that, the question arises, how to evaluate deformations according to the Second Order Theory, where the principle of the superposition of internal forces is not valid any more.
- quasi-permanent live loads and only one creep coefficient dependent on the material and the surrounding conditions: In this concept only one creep coefficient exists. The influence of the duration of load can be described by splitting up the live load into a quasi-permanent load, which causes creep deformations and a short-term load, which acts so short, that no creep deformation appears (comp. [9]).
    - quasi-permanent part of the live load

$$p_{\text{live load, perm.}} = \psi \cdot p_{\text{live load}} \tag{8.1}$$

    - short-term part of the live load

$$p_{\text{live load, short-term}} = (1 - \psi) \cdot p_{\text{live load}} \tag{8.2}$$

8.2 Influence of the duration of load in constant climate

For the determination of the influence of a temporary acting load, the model B according to Toratti [51] and the modified model B are subjected to the scenarios of the load history given in Fig. 8.1. In Fig. 8.2 and Fig. 8.3, respectively, the quasi-permanent part of a live load ψ according to Eq. (8.1) is given. As shown in these figures, the influence of the load decreases with decreasing duration of load. If the load is applied at the beginning of the load history (see Fig. 8.2(b)), the influence after 50 years can only be recognized if the duration

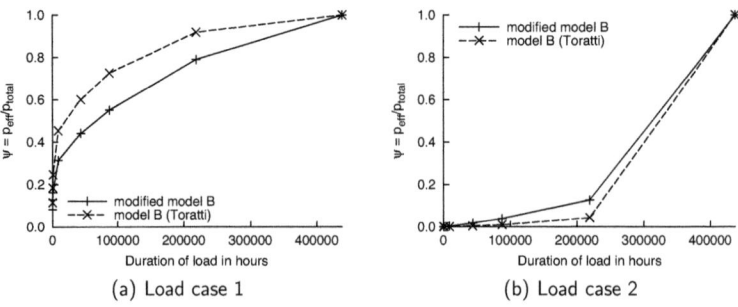

(a) Load case 1   (b) Load case 2

Figure 8.2: Influence of a live load once in 50 years (cases 1 and 2) on the creep coefficient
p_{eff} = quasi-permanent part of the live load
p_{total} = total live load

8.2 Influence of the duration of load in constant climate

of load has been longer than 25 years. If the load is applied at the end of the regarded time period, the duration of load influences the deflection even for short periods (see Fig. 8.2(a)).

For an annually repeating load, a nearly linear relation between annual duration of load and coefficient ψ can be found (see Fig. 8.3), so the creep deformation increases with increasing duration of load.

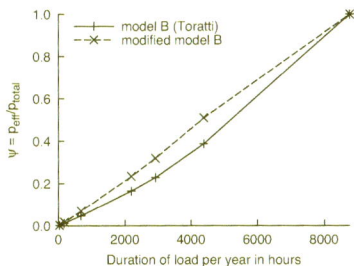

Figure 8.3: Influence of a repeating load (case 3, e.g snow load) on the creep coefficient in constant climate

In constant climate the creep strain caused by a constant stress can be evaluated by

$$\varepsilon_{cr} = \frac{\sigma}{E} \cdot \varphi = \frac{\sigma}{E} \cdot m \cdot t^c \tag{8.3}$$

where ε_{cr} creep strain
  σ stress
  E Modulus of Elasticity
  φ creep coefficient
  m, c coefficients
  t time

Since the creep strain is superposed of the creep strains caused by different stresses, the strains after unloading can be evaluated by following equation

$$\varepsilon = \frac{\sigma}{E} \cdot m \cdot ((t - t_0)^c - (t - t_0 - \Delta t)^c) \tag{8.4}$$

where Δt duration of loading
  t_0 time of loading

For several loadings and unloadings the strain can be determined to be

$$\varepsilon = \sum_{i=0}^{n} \frac{\sigma_i}{E} \cdot m \cdot ((t - t_{0,i})^c - (t - t_{0,i} - \Delta t_i)^c) \tag{8.5}$$

where n numbers of loading and unloading

Since in these cases only annually repeating loads are of interest, the time of loading can be expressed as a function of the numbers of loadings and unloadings

$$\varepsilon = \sum_{i=0}^{n} \frac{\sigma_i}{E} \cdot m \cdot ((t - i \cdot \Delta t_y)^c - (t - i \cdot \Delta t_y - \Delta t_i)^c) \tag{8.6}$$

where Δt_y duration of an interval, e.g. a year
This equation can be simplified to an effective creep coefficient of the system

$$\varphi_{ef} = m \cdot \sum_{i=0}^{n} ((t - i \cdot \Delta t_y)^c - (t - i \cdot \Delta t_y - \Delta t_i)^c) \tag{8.7}$$

Since the coefficient ψ can be also interpreted as a reduction of the creep coefficient

$$\varepsilon_{cr} = \frac{\sigma}{E} \cdot \underbrace{\psi \cdot \varphi}_{\varphi_{eff}} \tag{8.8}$$

the coefficient can be determined to be

$$\begin{aligned}\psi &= \frac{\varphi_{ef}}{\varphi} \\ &= \sum_{i=0}^{n} \left(1 - \frac{i \cdot \Delta t_y + \Delta t_{St}}{t}\right)^c - \left(1 - \frac{i \cdot \Delta t_y + \Delta t_{DOL} + \Delta t_{St}}{t}\right)^c\end{aligned} \tag{8.9}$$

where
- t current time
- Δt_y duration of a cycle
- Δt_{DOL} duration of the load during a cycle
- Δt_{St} starting point
- c exponent of the creep function (see Tab. 7.1)
  $= 0.21$ for Toratti [51]'s model B
  $= 0.35$ for the modified model B

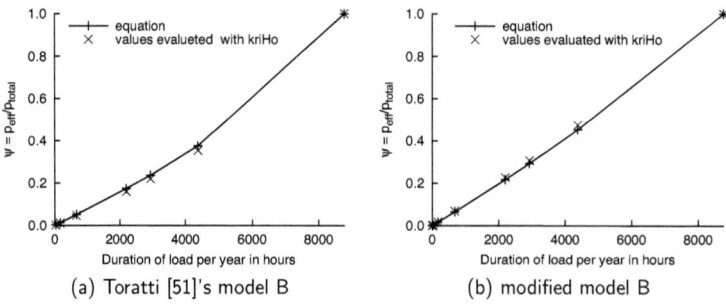

Figure 8.4: Comparison between the coefficient ψ evaluated by *kriHo* and using Eq. (8.9)

Fig. 8.4 shows a comparison of the effective part of the live load ψ in constant climate evaluated by *kriHo* and with Eq. (8.9). As can be seen, there is sufficient correspondence between the values evaluated by both methods. So Eq. (8.9) can be used for the determination of

8.2 Influence of the duration of load in constant climate

the quasi-permanent part of a live load.

Applying the equations on the scenarios given in Fig. 8.5, the quasi-permanent part of an

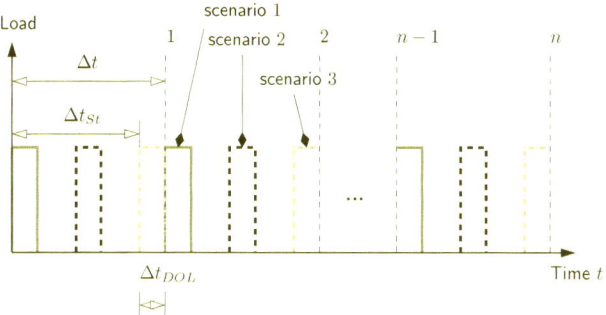

Figure 8.5: Load scenarios

annually repeating live load can be determined by the following equations (see Fig. 8.6)

$$\psi_{min} = m \cdot t_{DOL} \tag{8.10}$$

$$\psi_{max} = \frac{1 - c_{DOL}}{8760 \frac{h}{year}} \cdot t_{DOL} + c_{DOL} \tag{8.11}$$

where t_{DOL} duration of load per year in hours
m, c_{DOL} coefficients given in Tab. 8.1

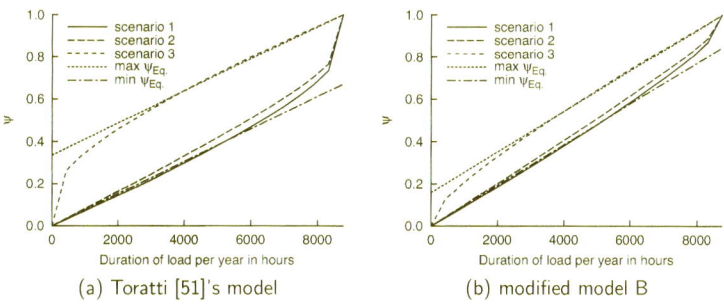

(a) Toratti [51]'s model   (b) modified model B

Figure 8.6: Maximum and minimum coefficient ψ, respectively, for an annually repeating load according to Eq. (8.9)

Table 8.1: Coefficients m and c_{DOL} for the determination of the quasi-permanent part of an annually repeating live load

	Toratti [51]'s model B	modified model B
m	7.65776E-05	9.64957E-05
c_{DOL}	0.335423977	0.159844344

8.3 Influence of an annually repeating load, e.g. snow in variable climate

Non-heated roof structures are normally loaded annually by a snow load. Since they are not heated, the relative humidity varies during the year. For the evaluation of the quasi-permanent part of the snow load, it is assumed, that the course of the relative humidity is affine to the average relative humidity of Löffingen (see Fig. 8.7), since measurements are performed in this region, in order to compare the evaluated results with the measured ones (see Sec. 6.3).

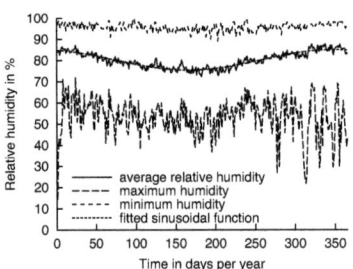

Figure 8.7: Climate in the region of Löffingen

Besides that, it is assumed, that the erection of the structure is on the 18.10. and it snows on the 01.01. every year. The annual course of the relative humidity as well as the course of the external load is given in Fig. 8.8.

Based on these assumptions, several cases are evaluated (see Fig. 8.9), varying the width of the cross section b, the average relative humidity RH and the annual amplitude of the relative humidity ΔRH.

As shown in Fig. 8.9, the quasi-permanent part of the load may even become larger than 100%. The increase of the quasi-permanent part can be explained by the explanation model according to Grossmann [19] (see Chap. 2.3). In this explanation model, the creep is caused by breaking of the hydrogen bonds between the cellulose chains in the timber. At the point in time of the assumed loading, the moisture content reaches nearly its maximum value (see Fig. 8.10). Therefore, only few hydrogen bonds between the single cellulose chains exist (see Fig. 8.11). The loading of the hydrogen bonds is on average higher than in a lower moisture content. Due to the loading some of the hydrogen bonds break, resulting in a

8.3 Influence of an annually repeating load, e.g. snow in variable climate

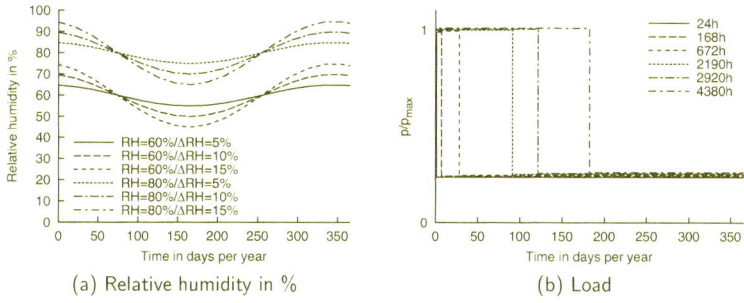

(a) Relative humidity in %

(b) Load

Figure 8.8: Considered annual loading and course of relative humidity

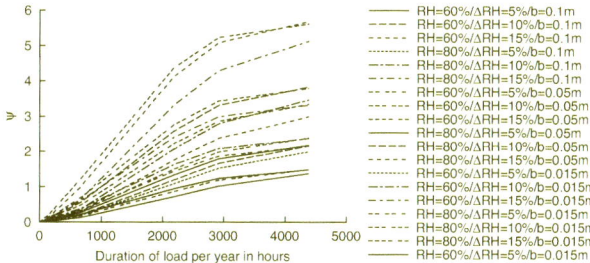

Figure 8.9: Ratio of the quasi-permanent load and permanent load ψ evaluated by *kriHo* using the modified model B for different average relative humidities RH, different amplitudes ΔRH and different cross section dimensions b

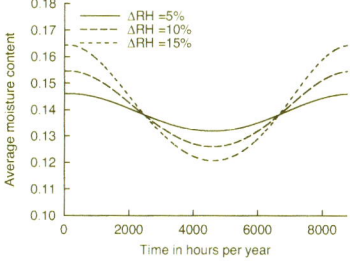

Figure 8.10: Assumed average moisture content for the climate given in Fig. 8.8(a) ($RH=60\%$)

movement of the single cellulose chains and therefore in a creep deformation. At the point in time of the unloading, the moisture content is lower than at the point in time of the loading. Therefore some hydrogen bonds are remade during drying. So in the deformed configuration,

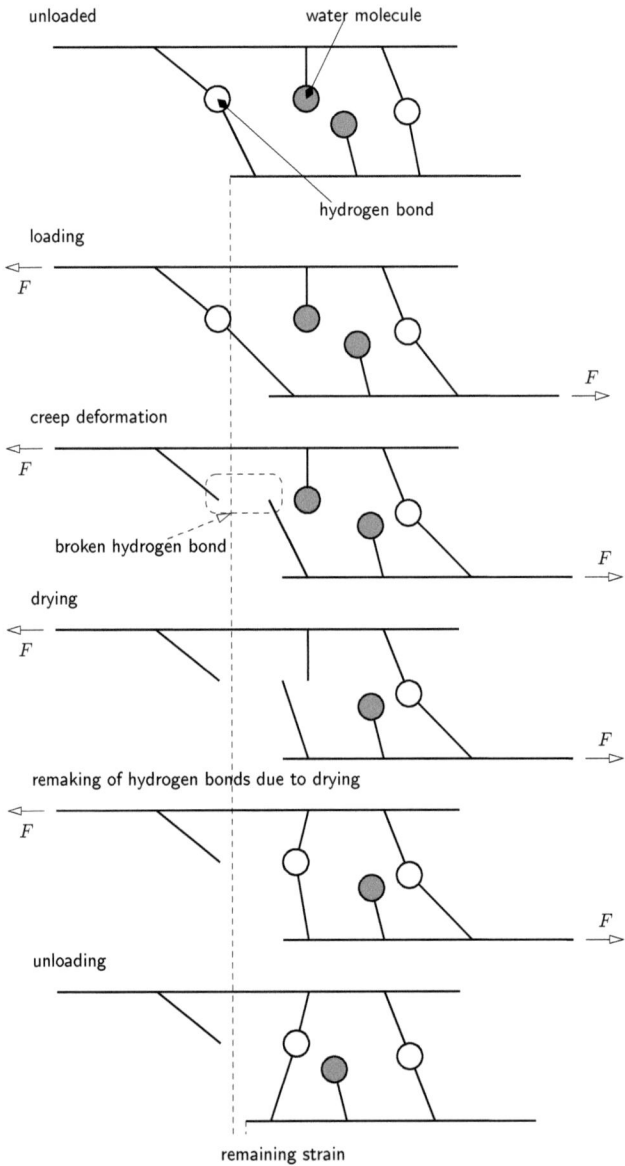

Figure 8.11: Schematic explanatory model of the remaining creep strain for an annually repeating load in variable moisture contents

8.3 Influence of an annually repeating load, e.g. snow in variable climate

more hydrogen bonds exist at the point of unloading than at the point of loading. Finally, no complete recovery can take place, since the hydrogen bonds, remade in the deformed configuration, prevent a movement of the cellulose chains in the original positions. So every loading and unloading with different moisture contents results in an additional creep strain, leading finally to ψ-coefficients larger than 1.0.

For the analytical determination of the coefficient ψ in variable climate, it is assumed, that a creep function as a power function comparable to Eq. (8.3) is valid. In contrast to Eq. (8.3) the Modulus of Elasticity is dependent on the current moisture content

$$\varepsilon_{cr} = \frac{\sigma}{E(u(t))} \cdot \varphi = \frac{\sigma}{E(u(t))} \cdot m \cdot t^c \qquad (8.12)$$

where
- ε_{cr} creep strain
- σ stress
- E Modulus of Elasticity
- φ creep coefficient
- m,c coefficients
- t time

Based on this equation, the coefficient ψ can be determined according to Sec. 8.2, resulting in the following equation

$$\begin{aligned}\psi =& k_{system} \cdot \sum_{i=0}^{n} \left(1 - \frac{i \cdot \Delta t_y + \Delta t_{St}}{t}\right)^c \\ &- k_{system} \cdot \sum_{i=0}^{n} \frac{E(u(t - \Delta t_{St}))}{E(u(t - \Delta t_{DOL} - \Delta t_{St}))} \cdot \left(1 - \frac{i \cdot \Delta t_y + \Delta t_{DOL} + \Delta t_{St}}{t}\right)^c\end{aligned} \qquad (8.13)$$

where
- t current time
- Δt_y duration of a cycle
- Δt_{DOL} duration of the load during a cycle
- Δt_{St} starting point
- c exponent of the creep function (see Tab. 7.1)
  - $= 0.21$ for Toratti [51]'s model B
  - $= 0.35$ for the modified model
- k_{system} system parameter according to Tab. 8.2
- E average Modulus of Elasticity

Table 8.2: System parameter

	k_{system}
Toratti [51]'s model B	1.15
modified model B	1.21

In Eq. (8.13) a system parameter k_{system} is introduced, because in the equation only the effects of normal creep are considered. However, the mechano-sorptive creep also influences the quasi-permanent part of the live load, which is considered by the k_{system} coefficient. As shown in Fig. 8.12, the evaluation with *kriHo* fits Eq. (8.13), as long as the duration of load per year is shorter than half a year.

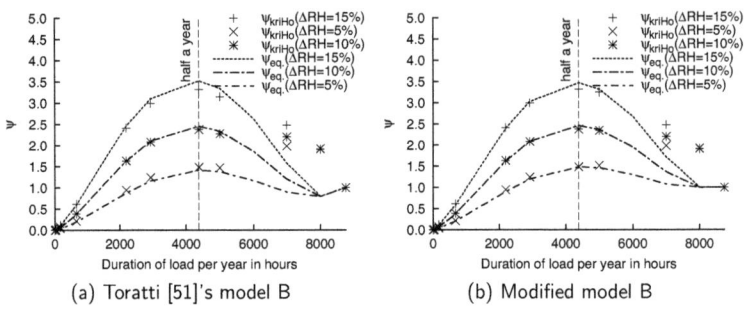

(a) Toratti [51]'s model B

(b) Modified model B

Figure 8.12: Comparison of the quasi-permanent part of the live load ψ in variable surrounding conditions according to Eq. (8.13) and evaluated by kriHo

8.4 Comparison to the measurements by Gutenkunst [20]

As already mentioned in Sec. 6.3, Gutenkunst [20] performed measurements on the time-dependent deflections in buildings, where the snow load obviously should not be neglected. Within this study, creep coefficients of about 2.75 to 5.7 are determined, in case the quasi-permanent part of the snow load is not taken into account.

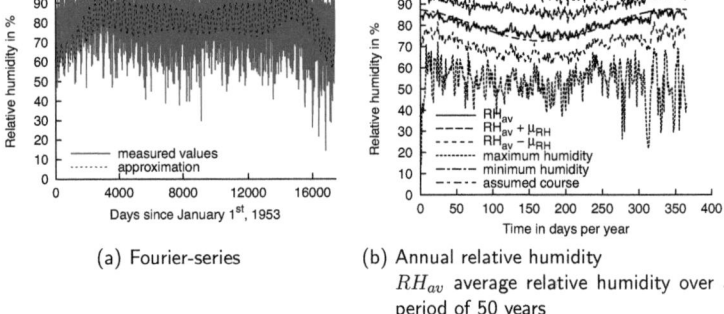

(a) Fourier-series

(b) Annual relative humidity
RH_{av} average relative humidity over a period of 50 years
μ_{RH} standard deviation of the relative humidity

Figure 8.13: Course of the relative humidity in Löffingen for the determination of the quasi-permanent part of the snow load

For the evaluation of the ψ-coefficient, the change of the relative humidity ΔRH influences the Modulus of Elasticity and therefore the ψ-coefficient. For the determination of the amplitude of the relative humidity, the relative humidity is approximated by a set of Fourier-series. If only the series with an amplitude larger than 1% are considered, the course given

8.4 Comparison to the measurements by Gutenkunst [20]

in Fig. 8.13(a) is obtained. The average ΔRH is determined by

$$\Delta RH = \sum^{n_{years}} \frac{RH_{max,FS,a} - RH_{min,FS,a}}{2 \cdot n_{years}} \tag{8.14}$$

where ΔRH amplitude of the relative humidity
$RH_{max,FS,a}$ annual maximum relative humidity according to the approximation by a set of Fourier series
$RH_{min,FS,a}$ annual minimum relative humidity according to the approximation by a set of Fourier series

In Tab. 8.3 the ψ-coefficient for the measured buildings in the region of Löffingen are evaluated.

For this evaluation, the relative humidity according to Fig. 8.13(b) are used. Based on this course of the relative humidity the moisture contents and the stiffness of the cross section can be determined for the evaluation of the effective part of the quasi-permanent part of the live load according to Eq. (8.13). So the effective creep coefficient of the element can be evaluated and compared to the measured ones.

As shown in Tab. 8.3 the effective creep coefficients related to the dead load evaluated by means of the effective part of the snow load according to Eq. (8.13) fit the measurements of the buildings Unadingen, whereas a larger difference exists in the comparison of the values of the buildings in Reiselfingen and Behla. However, the evaluation is based on average values as e.g. average duration of snow, average snow load etc. are considered. Therefore, differences between the real courses of these values and the assumed courses are supposed to lead to the differences in the determination of the quasi-permanent part of a live load. Nevertheless at the time being, the differences of 20% for the time dependent deformation of the evaluation and the measurements seem to be acceptable. Therefore the quasi-permanent part of the dead load in variable climate can be approximated by Eq. (8.13) in constant as well as in variable climate.

Table 8.3: Comparison between the effective part of the snow load $\psi \cdot s$ in the measurements according to Gutenkunst [20] and according to Eq. (8.13)

	Behla	Unadingen	Reiselfingen
year of construction	1968	1977	1971
age	40	31	37
h in m	0.135	0.15	0.135
b in m	0.095	0.075	0.075
distance of the rafters in m	0.74	0.625	0.765
dead load in kN/m²	0.2496	0.45725	0.459396
snow load in kN/m²	0.296	0.25	0.306
duration of load in hours	2016	2016	2016
$RH_{average}$ in %	80.16	80.16	80.16
ΔRH in % (see Fig. 8.13)	7.5	7.5	7.5
$u_{eq}(t_{unloading})$ according to Eq. (3.21) in %	18.7	18.7	18.7
$u_{eq}(t_{loading})$ according to Eq. (3.21) in %	21.4	21.4	21.4
$\Delta u_{surface}$ in %	2.7	2.7	2.7
Δu_{min} according to Eq. (7.12) in %	1.6	2.1	2.1
E_0 in kN/cm²	1196.22	1229.25	515.34
$E_{loading}$ according to Eq. (3.17) in kN/cm²	1170.15	1195.72	501.29
$E_{unloading}$ according to Eq. (3.17) in kN/cm²	1196.22	1229.25	515.34
φ_0	2.23	2.23	2.23
ψ according to Eq. (8.13)	0.79	0.76	0.91
$\varphi_{evaluated}$[a] (modified model B)	4.32	3.16	3.58
$\varphi_{measured}$	5.78	3.21	2.75
$\varphi_{evaluated}/\varphi_{measured}$	75%	99%	130%
$w_{evaluated}/w_{measured}$[b]	79%	99%	122%

[a] determination of $\varphi_{evaluated}$

$$\varphi_{evaluated} = \frac{g \cdot \varphi + s \cdot \psi \cdot \varphi}{g}$$

where g deadload
        s snow load
        φ creep coefficient
             =2.23 (see Sec. 5.3)

[b] w = midspan deflection

9 Effects of the increased creep deformations on the behavior of systems

9.1 Influences on the ultimate resistance of columns

9.1.1 General

The load capacity of a column is reached, when either the stresses reach the ultimate strength or the deformation of the column is so large, that no equilibrium of forces can be found. The load capacity depends on the initial imperfection of the column. Since the creep strain is a strain without a stress, the creep strains may be interpreted as an additional geometrical imperfection.

For this reason, a critical ratio between permanent and total load is given in DIN 1052 [9], for which the creep deformations have to be considered (see Becker [2] and Hartnack [24][1]). This limit is set at 70%, which means that the creep deformations have to be considered in the design of columns, if the permanent load represents 70% of the total load. These rules are linked to the creep coefficients given in DIN 1052 [9]. For service class 2 a creep coefficient of 0.8 is used.

However, since the creep coefficient determined in the measurements is larger than the values given in DIN 1052 [9], it is questionable, whether this limit of the ratio between the permanent load and the total load for the consideration of creep is still valid.

9.1.2 Determination of the critical ratio of the permanent load and the total load based on DIN 1052 [9] and based on the modified creep coefficient

If the creep deformation is only considered in cases where the permanent load is 70% or more of the total load – as proposed in DIN 1052 [9] –, a certain exceeding of the assumed load capacity is accepted, in case, the permanent load is below this limit, because even below this limit of the ratio between permanent and total load, creep deformations occur. Therefore the imperfection of the system increase during the time compared to the assumed and considered imperfection at a time $t = 0$ resulting in larger stresses than evaluated.

In order to determine the critical ratio of permanent load and total load, for which the creep

---

[1]Becker [2] and Hartnack [24] put a lot of work in discussing this question. Within the scope of this study, it is hardly possible to discuss all the parameters as both did. For this reason this chapter just shows tendencies and a rough approximation on the effect of larger creep coefficients on columns.

deformations have to be considered, the following procedure is used:
- The studied system is a column with a support representing Euler-case II (see Fig. 9.1). In this case, the deformation with respect to the creep deformation can be determined

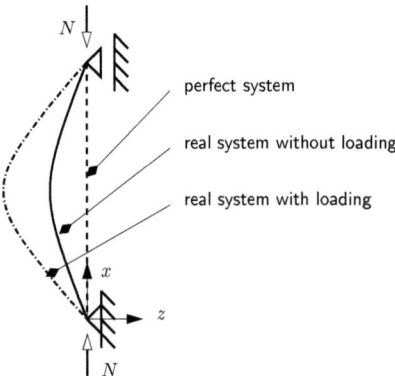

Figure 9.1: System studied for the effects of the increased creep deformations

by the following equations
- differential equation

$$EJ \cdot w''_{el} + N \cdot (w_{el} + w_0) = 0 \qquad (9.1)$$

where w_{el} elastic deformation
       w_0 imperfection
       P external axial force
       EJ bending stiffness

- solution of the differential equation, assuming a sinusoidal imperfection w_0 of the system

$$w_{el}(x) = w_{el,m} \cdot sin\left(\frac{\pi}{L} \cdot x\right) \qquad (9.2)$$

$$w_{el,m} = -\frac{N \cdot w_0 \cdot L^2}{EJ \cdot \pi^2 + N \cdot L^2 \cdot (1 + k_{def})} \qquad (9.3)$$

where $w_{el}(x)$ elastic deformation at a location x of the column
       $w_{el,m}$ elastic deformation at midspan of the column
       w_0 imperfection at midspan
       L length of the column
       EJ bending stiffness of the column
       N normal force in the column

- determination of the ultimate resistance of a column (Euler-case II): For the evaluation of the load capacity of the column, the design proposal according to DIN 1052 [9] for the evaluation of Second-Order Theory is used. Within this procedure it is assumed, that the ratio of permanent load and total load is lower than 70%. In this case the creep deformation need not be considered according to DIN 1052 [9] ($\rightarrow k_{def} = 0$)

9.1 Influences on the ultimate resistance of columns

The bending moment in the column can be evaluated by

$$M(x) = -EJ \cdot \frac{d^2 w_{el}(x)}{dx^2} = EJ \cdot w_{el,m} \cdot \left(\frac{\pi}{L}\right)^2 sin\left(\frac{\pi}{L} \cdot x\right) \qquad (9.4)$$

$$M(x = L/2) = EJ \cdot w_{el,m} \cdot \left(\frac{\pi}{L}\right)^2 \qquad (9.5)$$

where $w_{el}(x)$ elastic deformation at location x of the column
$w_{el,m}$ elastic deformation in midspan of the column
w_0 imperfection at midspan
L length of the column
EJ bending stiffness of the column
N normal force in the column
M bending moment according to Second-Order Theorie

For the structural design of the column the failure criterion is given by

$$1 = \frac{\frac{N}{A}}{f_{c,0,d}} + \frac{\frac{M}{W}}{f_{m,d}} \qquad (9.6)$$

- determination of the permanent load: Solving Eq. (9.6), the maximum normal force of the assumed column can be determined. This total load consists of the permanent load as well as of the live load. Since in this procedure no creep deformation is considered, the maximum part of the permanent normal force is 69.9% $\simeq$ 70% of the total load.

$$N_{\text{permanent}} = 70\% \cdot N_{\text{total}} \qquad (9.7)$$

- determination of the creep deformations: The creep deformation due to the permanent load is determined according to Eq. (9.3)

$$w_{el} = f(N_{\text{permanent}}, k_{def} = 0.8) \qquad (9.8)$$
$$w_{cr} = k_{def} \cdot w_{el} \qquad (9.9)$$

- In order to consider the creep deformation, the imperfection is increased by the creep deformation.

$$w_0 = w_0 + w_{cr} \qquad (9.10)$$

The idea behind this consideration of the creep strain is, that a column is loaded during the time period in question. At the end of this period, the column is unloaded. At this stage, the remaining deformation at midspan of the column consists of the initial imperfection at time $t = 0$ and the time dependent deformation due to creep. If the column is loaded again, the existing midspan deformation has to be considered as an additional imperfection of the column for the determination of the internal forces.

- determination of the stresses and the utilization u of the column with respect to the increased imperfection: Within this step, the "real" internal forces are evaluated, in order to determine the accepted exceeding of the stresses, if the rules of DIN 1052 [9] are applied on systems with a ratio of permanent load vs. total load $\leq 69.9\% \simeq 70\%$.

$$u_{acceptable} = \frac{\frac{N_{\text{total}}}{A}}{f_{c,0,d}} + \frac{\frac{M}{W}}{f_{m,d}} \qquad (9.11)$$

- determination of the permanent load by a numerical solution of Eq. (9.8) to Eq. (9.11) in order to fulfill the following boundary conditions
    - equal exceeding of the maximum stresses due to the common rules using the creep coefficient of $k_{def} = 0.8$ as proposed in DIN 1052 [9] and the modified creep coefficient

$$u_{\text{accepted}}(k_{def} = 0.8) \equiv u_{\text{accepted}}(k_{def,\text{modified}}) \tag{9.12}$$

    - evaluation of the total load

$$N_{\text{permanent}} + N_{\text{short-term}} = N_{\text{Load capacity without creep}} \tag{9.13}$$

If this procedure is applied to given geometries and material properties, the critical limit can be evaluated, at which the creep deformations have to be considered in the structural design of columns, achieving the same level of acceptance as in the current standard DIN 1052 [9]. For the numerical determination of this critical ratio, the cases given in Tab. 9.1 are studied. In Fig. 9.2 the ratio $N_{\text{permanent}}/N_{\text{total}}$ for which the same exceeding of the utilization according

Table 9.1: Range of parameters for the determination of the critical ration $N_{\text{permanent}}/N_{\text{total}}$

	lower boundary	upper boundary
material	C24 / GL24h	C30 / GL28h
length	2000mm	10000mm
breadth × height	100mm × 100mm	340mm × 340mm

to the current rules for different creep coefficients and different material properties are given. Fitting a function to the results given in Fig. 9.2 by the method of least squares, the critical ratio can be determined by

$$\frac{N_{\text{permanent}}}{N_{\text{total}}} = a \cdot \varphi^b + \lambda \cdot c \cdot (\varphi - \varphi_0)^d \tag{9.14}$$

where φ creep coefficient
a,b,c,d coefficients in dependence on the material according to Tab. 9.2
φ_0 creep coefficient in service class 2 according to DIN 1052 [9]
$= 0.8$

This means, that for the region of Tübingen with a creep coefficient of $\varphi = 2.23$ the critical ratio is

$$\frac{N_{\text{permanent}}}{N_{\text{total}}} = 0.0016276 \cdot \lambda + 0.221408979 \tag{9.15}$$

where λ slenderness of the column

A comparable reduction of the acceptable ratio of the permanent load and the total load is expected for lateral torsional buckling (see Kuhlmann and Teichmann [34], Kuhlmann and Hofmann [33] and Hofmann [26]).

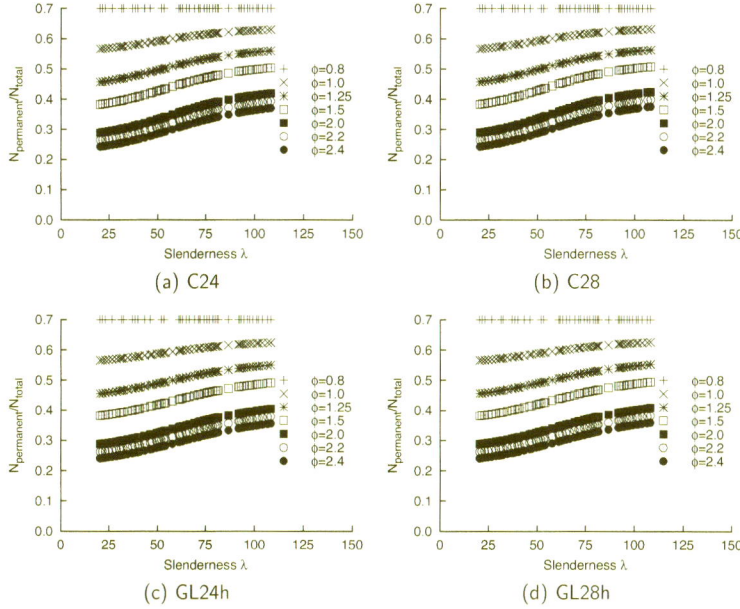

Figure 9.2: Critical ratio $N_{\text{permanent}}/N_{\text{total}}$, at which creep deformations should be considered in order to achieve the same level of acceptance as proposed in DIN 1052 [9], depending on the creep coefficient and the slenderness of the column

Table 9.2: Parameters for Eq. (9.14)

Material	a	b	c	d
GL24h	0.548715563	-1.122803252	0.001327003	0.310179815
GL28h	0.548755319	-1.124195344	0.001372545	0.309090294
C24	0.549588149	-1.128772745	0.001504864	0.293722252
C28	0.548488789	-1.152304166	0.001634212	0.304973372
average value	0.548886955	-1.132018877	0.001459656	0.304491433

9.2 Influence on timber-concrete composite slabs

Within the timber concrete composite slabs or, more generally spoken, in composite slabs made of elements with different time-dependent behavior, the time-dependent behavior directly influences the stresses and therefore the ultimate limit state.

In difference to steel-concrete composite structures, timber can not yield comparably to steel. Therefore, the stresses with respect to the time dependent deformations have to be considered in every critical point in time. Due to the different development of the creep strain and shrinkage of timber and concrete, it is proposed to consider at least three different points in time in the structural design of timber concrete composite structures (see [48]):

- directly after erection $t = 0$
- within the range of time between 3 and 7 years
- at the final point in time, normally after 50 years

At time $t = 0$ the elastic stress distribution takes place. Therefore, the normal forces in the composite elements as well as the shear forces reach often their maximum values. Until the period of time between 3 to 7 years, the concrete creeps faster than the timber. Additionally, the concrete finalizes its shrinkage. Both effects lead to a maximum loading of the timber since the concrete withdraws its loading. This point in time becomes the more critical, the slower the timber is creeping. For a general proposal the course of the normal creep should be considered, in order to receive results on the safe side, since the normal creep exists independent of the surrounding conditions. At time $t = 50$ years both materials as well as the connection reach their maximum creep strain, so the deflection reaches the maximum value.

If the real creep coefficient of timber is larger than the assumed creep coefficients according to the standards the following effects are expected:

- $t = 0$: No effect is expected since creep does not take place yet.
- $t = 3$ to 7 years: No difference to the common design is expected, since the models are validated until this period of time. Therefore, the strains proposed by the models at that point in time are not influenced by the increased, measured creep coefficients.
- $t = 50$ years: At that point in time, the usually assumed creep coefficient is smaller than the measured ones. Therefore, the deflection will be underestimated in the common design.

Although the end creep values of timber and concrete are comparable, the stress redistributions at the critical point in time concerning the maximum stresses in timber have to be determined. For this determination the use of effective creep coefficients in pure concrete-concrete composite beams are proposed in Ruesch and Jungwirth [47]. Kreuzinger [32] extended this theory considering the deformability of the connection in the joint between timber and concrete. The consideration of the different temporal development of the creep strain in the design methods is given in [48]. Within this study, the different temporal development is split up in several intervals with an affine development of the creep coefficient of timber and concrete, in order to allow the solution of the differential equation of the effective creep coefficient according to Ruesch and Jungwirth [47] and Kreuzinger [32].

In Fig. 9.3 the ratio of the current creep coefficient and the creep coefficient of concrete after 50 years is set into dependence on the ratio of the current creep coefficient and the creep coefficient of timber after 50 years for the original model B according to Toratti [51], the model according to Hanhijärvi [22] and the modified model B. As can be seen, the temporal development of the creep coefficient of the modified model differs resulting in stress redistribution in the cross section.

Due to the increased creep coefficient of the measurements compared to the creep coefficients used in [48], the intervals for the temporal development of the creep coefficient has to be modified to the values given in Tab. 9.3 (see Fig. 9.3).

Within one interval i the effective creep coefficient in timber-concrete composite structures

9.2 Influence on timber-concrete composite slabs

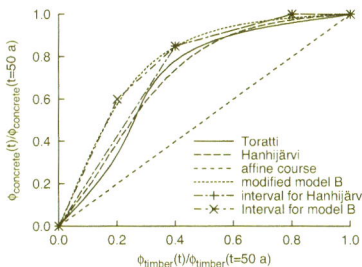

Figure 9.3: Temporal development of the creep coefficient of concrete dependent on the creep coefficient of timber

Table 9.3: Modified intervals for the determination of the effective creep coefficients in timber-concrete composite according to [48] due to the increased creep coefficient of the modified model B

interval	timber $\Delta\varphi_t/\varphi_t(t=50y)$	concrete $\Delta\varphi_c/\varphi_c(t=50y)$
Former intervals (see [48])		
1	40%	85%
2	40%	15%
3	20%	0%
Modified intervals		
1	20%	60%
2	20%	25%
3	40%	15%
4	20%	0%

can be determined by the following equation:

$$\Delta\varphi_{w,C,i} = \Delta\varphi_{w,M,i} \cdot \left(\frac{1 + \Delta\varphi_{u,M,i}}{\Delta\varphi_{u,M,i} - e^{-\Delta\psi_i} \cdot (\Delta\varphi_{u,M,i} - \Delta\psi_i)} - \frac{1}{\Delta\psi_i} \right) \quad (9.16)$$

where $\Delta\varphi_{w,V,i}$ effective composite creep coefficient of the component w in the interval i
  $\Delta\varphi_{u,M,i}$ material creep coefficient of the component u in the interval i
  i interval according to Tab. 9.3
  u,w component of the composite slab, e.g. timber or concrete
    $u \neq w$
  $\Delta\psi_i$ system creep coefficient according to Eq. (9.17)

One input parameter in this equation is the system creep coefficient, which can be determined by

$$\Delta\psi_i = \frac{\delta_u \cdot \Delta\varphi_{u,M,i} + \delta_w \cdot \Delta\varphi_{w,M,i}}{\delta_u + \delta_w} \tag{9.17}$$

where δ_u flexibility of the component u
$\Delta\varphi_{u,M,i}$ material creep coefficient of the component u in the interval i
u,w component of the composite slab, e.g. timber or concrete
$u \neq w$

The total creep coefficient of the component u in timber-concrete composite structures can be determined as the sum over all integrals

$$\varphi_{u,C} = \sum_{i=1}^{4} \Delta\varphi_{u,C,i} \tag{9.18}$$

For a more detailed discussion about the influence of the time-dependent behavior in timber-concrete composite slabs see among others Fragiacomo [15], [48] and [49].

9.3 Conclusions

In the common structural design, the creep deformation influences – among others – the ultimate resistance of columns and the stress distribution in composite systems. For the design of columns according to the current standards, the creep deformation have to considered if the ratio between the permanent load and the total load exceeds 70%. Since the creep coefficients in the standards (see e.g. DIN 1052 [9]) are lower than the measured creep coefficients, this ratio drops to 22% to 46% in dependence on the slenderness of the column.

For the evaluation of the stresses in a timber-concrete-composite system, effective creep coefficients have to be determined. For this reason, the temporal development of the creep coefficient has to be split up in several intervals. Within this interval, an affine development can be assumed, so the effective creep coefficient can be evaluated according to Ruesch and Jungwirth [47] and Kreuzinger [32]. Since the course of the creep strain and the absolute creep coefficient after 50 years is modified, the intervals also have to be adapted in order to consider the larger creep coefficients in the design of composite slabs.

With these modifications, the effects of the increased creep coefficients obtained by measurements can be considered in the structural design of columns and timber-concrete-composite slabs.

10 Conclusion and Outlook

The time-dependent behavior of timber influences the ultimate limit state as well as the serviceability limit state of structures. For the consideration of the time-dependent behavior creep coefficients have been introduced into the standards. However, these creep coefficients do not consider all important influences as e.g. variation of the moisture. In order to consider the main parameters for the time dependent behavior of timber, different rheological models have been developed by various researchers.

Within this study the rheological models according to Toratti [51], Hanhijärvi [22], Becker [2] and Mårtensson [38] are compared. The comparison shows, that the models lead to different time-dependent deformations if they are extrapolated over the period of 50 years. For the same situation, assuming a constant climate, the creep coefficient can reach values between 0.52 and 1.1 in dependence on the model used. In variable climate the creep coefficients vary between ≈ 0.9 to ≈ 1.8. Besides that various influences are considered in different ways. For constant surrounding conditions increasing initial moisture content leads to a decrease of the creep coefficients evaluated by the models according to Toratti [51] and Becker [2], whereas in the model according to Hanhijärvi [22] the initial moisture content hardly influences the creep coefficients after 50 years. This model is the only one of the studied models which consists of elements in parallel order. Therefore the absolute creep strain is limited, independent of the reason for the creep, so there are hardly differences in the results after a period of 50 years, no matter whether the moisture content varies or not.

In most of the models, the creep strain depends linearly on the strain, except in the model according to Mårtensson [38] and Hanhijärvi [22]. The dependence of the creep coefficient according to Hanhijärvi [22]'s model on the Modulus of Elasticity is quite small for a period of 50 years. In the model according to Mårtensson [38], an obvious dependence of the creep coefficient on the Modulus of Elasticity exists.

In order to discuss the influence of the variability of Modulus of Elasticity on the creep coefficients, the Modulus of Elasticity is statistically distributed over the cross section according to the values given in [30] and the creep deformations are eveluted using the model according to Mårtensson [38]. Within this study, it can be shown, that the resulting creep coefficient of the beam only depends on the average Modulus of Elasticity. Therefore the influence of the variability of the Modulus of Elasticity on the creep coefficient may be neglected for practical purposes.

Since the creep coefficients of the models differ, they should be compared to test results. However, it is hardly possible to perform tests over the interesting period of time of about 50 years. For this reason, the current deformation of buildings, erected about 50 years ago in the region of Tübingen, South-West Germany, have been measured. Additionally, the current elastic stiffness is determined by a test loading. With these two values the creep coefficients can be determined, which should have been used by the engineer in order to predict the deformation existing after 50 years.

As a result an average creep coefficient of 2.23 based on 56 elements can be determined.

However, the standard deviation of about 0.97 for the creep coefficients is quite large. Comparing these creep coefficients to the values evaluated by the rheological models, Toratti [51]'s model B fits the results best. As underestimating the creep coefficients, the model B according to Toratti [51] is modified so, that a better agreement between the measurements and the evaluated deformations is achieved, assuming that only the normal creep and not the mechano-sorptive creep have to be modified. Within this modification, only the Kelvin-Voigt-bodies with a large retardation time have been adapted, so that significant differences between the original and the modified model appear for a period of time larger than 20 years. Therefore, all validations of the original model are still valid.

In order to verify the modified model, measurements performed by Gutenkunst [20] on buildings in the region of Breisgau-Hochschwarzwald, South-West Germany are evaluated by the modified model. In this region a higher snow load and a longer duration of the snow load compared to the region of Tübingen can be found. As shown in the comparison of the measurements and the evaluation, the modified model also may be applied for this region.

For the simplified determination of the creep coefficients based on the rheological models, functions are adapted to the numerical results. In order to determine the time-dependent behavior a basic creep function is determined for each studied model first. In order to consider the effect of drying the coefficient DSF is introduced, which takes into account the mechano-sorptive creep due to drying. For the mechano-sorptive creep due to annually changing surrounding conditions, a function GMS is fitted to the numerical results. This function modifies the basic creep coefficient in order to consider the mechano-sorptive creep strain. For the interaction between the changing moisture content due to drying and due to the changing surrounding conditions, the drying process is transformed into an annually changing relative humidity.

Since it is expected, that a larger creep strain exists in the outer layers due to the larger moisture variations, effective cross section dimensions are determined. However, – in difference to the other models – some kind of mechano-sorptive creep limit is reached for quite small variations of the relative humidity for a period of 50 years in the model B according to Toratti [51] and the modified model B. For this reason, no effective dimensions are necessary in these models.

Creep deformations influence the serviceability limit state as well as the ultimate limit state, especially when dealing with composite systems or elements subjected to compression. For this reason the impact of the increased creep coefficient on the design of columns is approximated in order to reach a comparable acceptance level as the German standard DIN 1052 [9]. It regularizes, that the creep strain has to be taken into account, if the ratio between the permanent load and the total load exceeds 70%. Using the increased creep coefficients determined by means of the measurements, the ratio between the permanent load and total load drops to values between 22% and 46% in dependence on the slenderness of the column.

For the design of composite slabs, composite creep coefficients have to be determined. These composite creep coefficients depend on the ratio of the temporal development of the creep strain of the composite partners (see [48]). For this reason, the intervals given in [48] have been modified in order to fit the evaluated creep coefficients or the results according to the modified model B, respectively.

However, in all studies except for the re-evaluation of the measured elements a sinusoidal course of the surrounding conditions as well as a constant load within a certain interval is assumed. So the question arises, whether this is a suitable approximation, or if the real course of the surrounding conditions has to be considered.

Besides that, there is still the unsolved question, why the modified model fit the measured values and some test results, but leads to larger differences, if the moisture content changes rapidly, even in those cases, where the accumulated moisture content is comparable. Therefore it should be clarified, whether the assumption, that the mechano-sorptive creep only depends on the accumulated moisture content, is valid in all cases. If not, the surrounding conditions should be determined, in which the assumption of this dependence leads to sufficiently exact results. Therefore the results of this study should only be directly applied to situations which are comparable to the situation in the region of Tübingen or the region of Breisgau-Hochschwarzwald, respectively. Unfortunately, the key parameters for defining this comparable situation are not yet clear. For this reason, additional measurements should be performed in regions, where the climate – especially the relative humidity and therefore the accumulated moisture content – differs significantly from the situations in Tübingen and Breisgau-Hochschwarzwald, as it is expected e.g. in the quite foggy regions of around the Lake of Konstanz or around the riverside of Donau. So the modified model can be compared to the results of these future measurements. If the model fits to the future results, the modification of the normal creep is verified. If the model does not fit to the future results, it is proposed to have a closer look on the modeling of the mechano-sorptive creep as it has been done within this study for the normal creep strain, since the mechano-sorptive should be the main differences between the single sets of measurements of the regions in Tübingen, around the Lake of Konstanz and around the riverside of Donau.

11 Bibliography

[1] J. Arfvidsson. Computer model for two-dimensional moisture transport. Manual for jam-2, Lund, 1989.

[2] P. Becker. *Modellierung des zeit- und feuchteabhängigen Materialverhaltens zur Untersuchung des Langzeitverhaltens von Druckstäben aus Holz*. PhD-thesis, Bauhaus-Universität Weimar, January 2002.

[3] E. Bou Saïd. *Contribution à la modelisation des effets différes du bois et du béton sous conditions climatiques variables. Application aux structures mixtes bois-béton*. PhD-thesis, 2003.

[4] J. D. Boyd. An anatomical explanation for visco-elastic and mechano-sorptive creep in wood and effects of loading rate on strength. In P. Baas, (Ed.), *New perspectives in wood anatomy*, pages 171–222. The Hague: Martinus Nijhoff and Dr. W. Junk Publishers, 1982.

[5] J. Carstensen. *Beiträge zum Biegekriechverhalten von Großquerschnitten unter baupraktischen Bedingungen*. PhD-thesis, Universität Hannover, IBH, 1993.

[6] Deutscher Wetterdienst. kl_dat_abgabe325_08. Technical report.

[7] F. Dieringer. *Dachsanierung des Kapellenbaus am Schloss Nymphenburg*. Diploma thesis, HTWG Konstanz, Prof. Dr.-Ing. W. Francke & Dr.-Ing. J. Schänzlin, January 2007.

[8] DIN 1052. *Holzbauwerke T.1. Berechnung und Ausführung und T. 2. Mechanische Verbindungsmittel mit Ergänzungen A1 und A2 (1996)*. 1988.

[9] DIN 1052. *Entwurf, Berechnung und Bemessung von Holzbauwerken - Allgemeine Bemessungsregeln für den Hochbau*. August 2004.

[10] DIN 1055. *Einwirkungen auf Tragwerke*. March 2001.

[11] DIN 1074. *Holzbrücken*. 1989.

[12] DIN 4074. *Sortierung von Nadelholz nach der Tragfähigkeit*. 2003.

[13] Eurocode 2. *DIN V ENV 1992 Eurocode 2: Planung von Stahlbeton- und Spannbetontragwerken; Teil 1: Grundlagen und Anwendungsregeln für den Hochbau*. June 1992.

[14] Eurocode 5. *DIN V ENV 1995 Eurocode 5: Entwurf, Berechnung und Bemessung von Holzbauwerken; Teil 1-1: Allgemeine Bemessungsregeln, Bemessungsregeln für den Hochbau*. June 1994.

[15] M. Fragiacomo. *Comportamento a lungo termine di travi composte legno-calcestruzzo.* PhD-thesis, Universität Trieste, 2000.

[16] M. Fragiacomo and J. Schänzlin. Modelling of timber-concrete floor structures. In A. Cecotti and S. Thelandersson, (Ed.), *Timber constructions in the new millenium.* Cost E5, September 2000.

[17] P. Gressel. Kriechverhalten von Holz und Holzwerkstoffen. *bauen mit holz*, pages 216–223, 1984.

[18] M. Grosse, R. Hartnack, S. Lehmann and K. Rautenstrauch. Modellierung von diskontinuierlich verbundenen Holz-Beton-Verbundkonstruktionen. *Bautechnik*, 80: 534 – 541 and 693 – 701, 2003.

[19] P. U. A. Grossmann. Mechano sorptive behaviour. In B. A. Boyd, J. A. Johnson and R. W. Perkins, (Ed.), *General constitutive relations of wood and wood-based materials*, pages 313–325, July 1978.

[20] M. Gutenkunst. *Bestimmung von effektiven Kriechbeiwerten durch Messungen an bestehenden Bauwerken.* Bachelor-thesis, HTWG Konstanz, Prof. Dr.-Ing. W. Francke & Dr.-Ing. J. Schänzlin, August 2008.

[21] M. Häglund. On moisture induced stresses in timber structural elements. COST E55 Meeting in Helsinki, March 2008.

[22] A. Hanhijärvi. *Modelling of creep deformation mechanisms in wood.* PhD-thesis, Helsinki University of Technology. Technical Research Centre of Finland. VTT Publications. Espoo (SF), 1995.

[23] A. Hanhijärvi. *Perpendicular-to-grain creep of Finnish softwoods in high temperature drying conditions – experiment and modelling in temperature range 95-125 °C.* VTT Publications 301, VTT Building Technology, 1997.

[24] R. Hartnack. *Langzeitverhalten von druckbeanspruchten Bauteilen aus Holz.* PhD-thesis, Bauhaus-Universität Weimar, April 2005.

[25] P. Hoffmeyer and R.W. Davidson. Mechano-sorptive creep mechanism of wood in compression and bending. *Wood Science and Technology*, 23: 215–227, 1989.

[26] R. Hofmann. *Erfassung des Langzeitverhaltens von schlanken Brettschichtholzträgern beim Stabilitätsnachweis "Kippen".* PhD thesis, Institut für Konstruktion und Entwurf, Universität Stuttgart, under preparation.

[27] Holzbautaschenbuch. *Holzbautaschenbuch*, volume 9. R. v. Haslasz; C. Scheer, 1995.

[28] R. J. Hoyle, R. Y. Hani and J. J. Eckhard. Creep of Douglas-fir beams due to cyclic humidity fluctuations. *Wood and Fibre Science*, 18: 468–477, 1986.

[29] D. G. Hunt. Linearity and non-linearity in mechano-sorptive creep in softwood in compression and bending. *Journals of material science*, 21: 2088–2096, 1986.

[30] Joint Committee on Structural Safety. *Probabilistic Model Code.* www.jcss.ethz.ch/, 2001.

11 Bibliography

[31] A. Kenel and U. Meierhofer. *Holz / Beton-Verbund unter langfristiger Beanspruchung.* Research report 115/39. EMPA Dübendorf (CH), Abteilung Holz, 1998.

[32] H. Kreuzinger. *Verbundkonstruktionen Holz / Beton.* 1994.

[33] U. Kuhlmann and R. Hofmann. Erfassung des Langzeitverhaltens von schlanken Brettschichtholzträgern beim Stabilitätsnachweis "Kippen". research report of DFG-project 542091, Institut für Konstruktion und Entwurf, Universität Stuttgart, 2009.

[34] U. Kuhlmann and G. Teichmann. Influence of creep on lateral torsional buckling of glued laminated timber beams. In *Proceedings of the 9th World Conference on Timber Engineering, WCTE 2006, Portland, OR, USA, August 6-10, 2006*, 2006.

[35] M. Leivo. *On stiffness changes in nail plate trusses.* PhD thesis, Espoo, Finland, 1991.

[36] A. Lühr. *Dachtragwerk der Münsterbauhütte Freiburg.* Bachelor-thesis, HTWG Konstanz, Prof. Dr.-Ing. W. Francke & Dr.-Ing. J. Schänzlin, August 2007.

[37] maps.google.de. Google Maps Deutschland. http://maps.google.de/maps?hl=de&tab=wl, 12.2008.

[38] A. Mårtensson. *Mechanical behaviour of wood exposed to humidity variations.* PhD-thesis, 1992.

[39] S. Mohager. *Studier av krypning hos trä.* KTH Stockholm, 1987.

[40] W. Moorkamp. *Zum Kriechverhalten hölzerner Biegeträger und Druckstäbe im Wechselklima.* PhD thesis, Universität Hannover, 2002.

[41] P. Morlier and L.C. Palka. Creep in Timber Structures. volume 8 of *Rilem Report*, pages 9 – 42. Rilem Publications, 1994.

[42] L. Muszynski, R. Lagana and S. M. Shaler. Characterization of hygro-mechanical properties of solid wood on the material level. In *Proceedings of the 8th World Conference on Timber Engineering, WCTE 2004, June 14-17, Lahti, Finland*, volume 2, pages 161–166, 2004.

[43] F.-H. Neuhaus. *Elastizitätszahlen von Fichtenholz in Abhängigkeit von der Holzfeuchtigkeit.* PhD-thesis, Institut für konstruktiven Ingenieurbau Ruhr-Universität Bochum, 1981.

[44] P. Niemz. *Physik des Holzes.* 2004.

[45] Österreichische Bergrettung. *Schnee- und Lawinenkunde.* Österreichische Bergrettung, Land Vorarlberg, www.bergrettung-koetschach.com/pdf/schnee_lawinenkunde.pdf, 11 2006.

[46] K. Rautenstrauch. *Untersuchungen zur Beurteilung des Kriechverhaltens von holzbiegeträgern.* PhD-thesis, Universität Hannover, IBH, 1989.

[47] H. Ruesch and D. Jungwirth. *Stahlbeton, Spannbeton - Berücksichtigung der Einflüsse von Kriechen und Schwinden auf das Verhalten der Tragwerke*, volume 2. Werner-Verlag, 1976.

[48] J. Schänzlin. *Zum Langzeitverhalten von Brettstapel-Beton-Verbunddecken*. PhD-thesis, Institut für Konstruktion und Entwurf, Universität Stuttgart, 2003.

[49] J. Schänzlin and M. Fragciacomo. Extension of EC5 Annex B formulas for the design of timber-concrete composite structures. CIB-40-10-1, International council for research and innovation in building and construction – working commission W18 - timber structures (CIB-W18), Meeting 40, Bled, Slovenija, http://www.rz.uni-karlsruhe.de/ gc20/IHB/cib.htm, August 2007.

[50] G. D. Taylor, D. J. West and B. O. Holsion. Creep of glued laminated timber under conditions of varying humidity. In *International timber engineering conference, London*, 1991.

[51] T. Toratti. *Creep of timber beams in variable environment*. PhD-thesis, Helsinki University of Technology, Laboratory of Structural Enginieering and Building Physics, 1992.

[52] P. Tukiainen and M. Hughes. Fracture of spruce and birch in the RT crack propagation direction. In *Proceedings of the 10th World Conference on Timber Engineering, WCTE 2008, Miyazaki, Japan, June 2-5, 2008*, 2008.

[53] wetter-online.de. Stuttgart-Echterdingen. webpage, http://www.wetteronline.de/cgi-bin/regframe?3&LANG=de&PLZ=70771&PLZN=Leinfelden-Echterdingen&PROG=citybild&PRG=citybild, 2008.

[54] wikipedia.de. Wikipedia. webpage, http://www.wikipedia.de/, 2008.

[55] Q. Wu and M. R. Milota. Rheological behaviour of douglas-fir perpendicular to the grain at elevated temperature. *Wood and Wood Fibre Science*, 27: 285–295, 1995.

A Verification of the implementation

A.1 General

In order to ensure that there are no errors concerning the implementation of the models, these models are compared to curves given in the description of the particular model. Therefore, the different models are often compared to systems with different boundaries and environmental conditions. However, the data are not always given in the necessary extend. So especially regarding the test set-up or the initial moisture content of the test specimen assumptions have to be made.

A.2 Model B according to Toratti [51]

The implementation of Toratti [51]'s model B is verified by the re-evaluation of the tests and evaluations, respectively, mentioned in Toratti [51] (see Fig. A.1). Within these tests a cross section of $h \times b = 89mm \times 89mm$ is loaded by a constant bending moment of 1.56kNm. During these tests, the climate varies between 40% and 90% within a 1 day and a 7 day cycle, respectively.

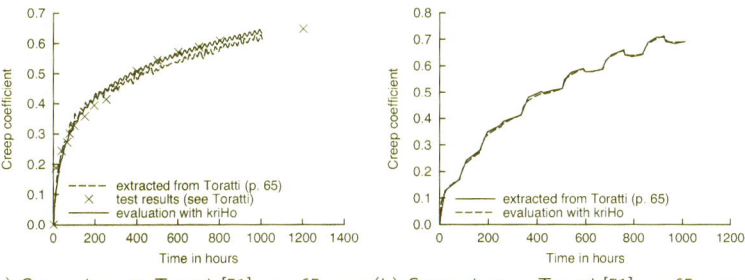

(a) Comparison to Toratti [51], p. 65, one day cycle

(b) Comparison to Toratti [51], p. 65, seven days cycle

Figure A.1: Comparison of the creep coefficient between the model B according to Toratti [51] and *kriHo*

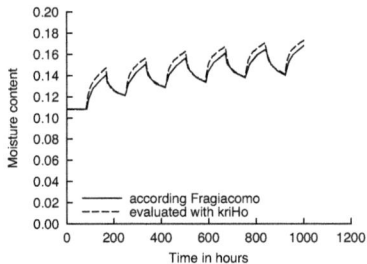

Figure A.2: Comparison of the moisture content between Fragiacomo [15] and *kriHo*

A.3 Model according to Hanhijärvi [22]

For the verification of the implementation of Hanhijärvi [22]'s model in *kriHo*, evaluations by Hanhijärvi [22] are re-evaluated by *kriHo*. In a first series, a bending beam is loaded on different stress levels. The climate is assumed to be constant. Since no values of the used cross section is given a timber cross section of h × b = 200mm × 100mm is assumed. The results of the evaluations are given in Fig. A.3.

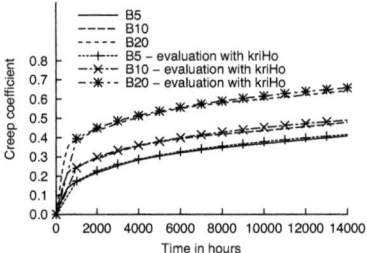

Figure A.3: Comparison of the creep coefficient between Hanhijärvi [22], p.113 and *kriHo* for constant moisture content

In order to verify the mechano-sorptive part of the implementation, bending tests cited by Hanhijärvi [22] are re-evaluated. In these tests the relative humidity varies between 35% and 90%. The cross section (h × b = 95mm × 45mm) is loaded on two different stress levels leading to a nominal edge stress of 8.2 MPa and 24.5 MPa, respectively.

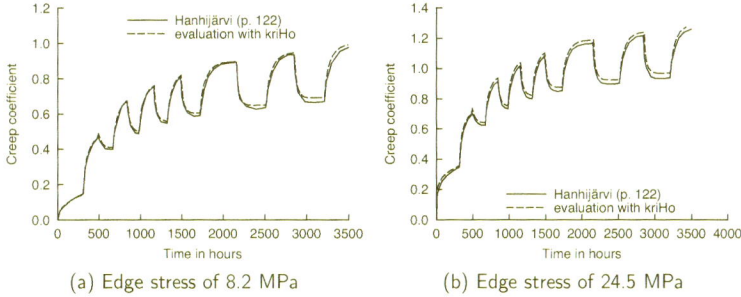

Figure A.4: Comparison of the creep coefficient between Hanhijärvi [22], p.122 and *kriHo* for changing moisture content and different bending moments

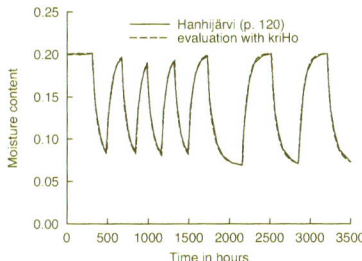

Figure A.5: Comparison of the moisture content between Hanhijärvi [22], p.120 and *kriHo*

A.4 Model according to Becker [2]

For linear creep deformation Becker [2] determines creep coefficients, given in Tab. A.1.

Table A.1: Creep coefficients for linear creep according to Becker [2]

Time	Time in h	creep coefficient
1 day	24	0.070
1 week	168	0.118
1 month	720	0.189
1 year	8760	0.414
10 years (end value)	87600	0.6

For the verification of the implementation of the mechano-sorptive part in the model according to Becker [2], tests cited in Becker [2] are re-evaluated (see Fig. A.7). Fig. A.7(a) shows the time dependent behavior of a beam (h × b = 95mm × 45mm), which is subjected to a variable climate between 35% and 90% in a 70 days-cycle. In Fig. A.7(b) the beam (h ×

A. Verification of the implementation

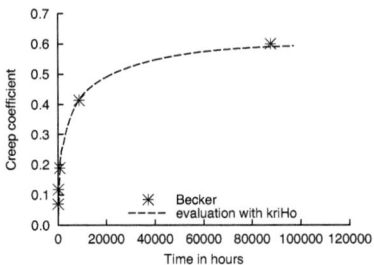

Figure A.6: Comparison of the creep coefficients due to linear creep between Becker [2] and *kriHo*

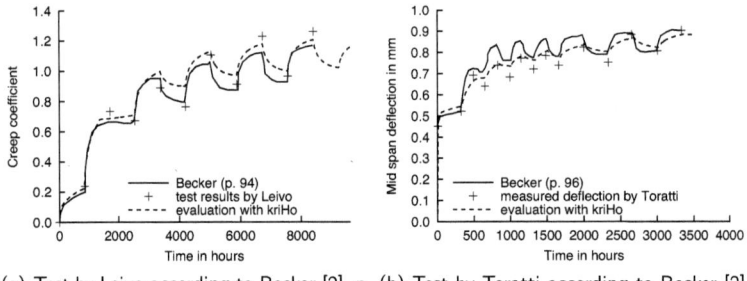

(a) Test by Leivo according to Becker [2], p. 94

(b) Test by Toratti according to Becker [2], p. 96

Figure A.7: Comparison between Becker [2] and *kriHo*

b = 95mm × 45mm) is subjected to a variable relative humidity, varying between 35% and 90%.

A.5 Model according to Mårtensson [38]

In Fig. A.8 the time dependent deformation of a beam (h × b = 95mm × 45mm) in variable climate is shown. As can be seen in Fig. A.8 *kriHo* and the results given by Mårtensson [38] fit

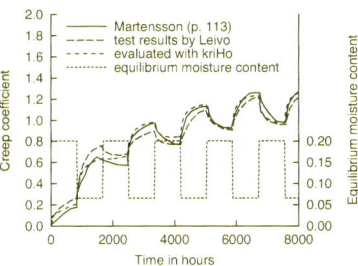

Figure A.8: Comparison between Mårtensson [38], p. 113 and *kriHo*

quite well, except for the first peak. One reason for these differences can be the determination of the moisture content. In Mårtensson [38] the moisture content is determined with the help of a tool developed by Arfvidsson [1], whereas in *kriHo* the diffusion coefficient and the surface emissivity according to the Toratti [51] are supposed to be valid.

B Bionic approach YaRM 0.1

B.1 General

In Chap. 6 a modification of the model according to Toratti [51] is shown. This modified model shows acceptable differences between the measured and the evaluated long term deformations. However, applying the model to the tests performed by Mohager [39] (see Toratti [51]), differences between the evaluation and the test results exist.

Comparing the key parameters of the tests and of the measurements concerning the time dependent behavior, one should expect larger creep deformations in the measurements than in the tests, since the accumulated moisture as well as the duration of the loads of the tests are smaller and shorter, respectively, than in the measured elements (see Tab. B.1).

Table B.1: Key parameters of the time dependent behaviour

	Tests by Mohager [39]	Measurements
Accumulated moisture	≈ 20	$>> 20$
Duration	≈ 1200 days	≈ 18250 days
Creep coefficient	≈ 3	≈ 2.3

However, the creep deformations of the tests are larger than the creep deformations obtained from measurements. Toratti [51] proposes model D for the re-evaluation of the tests performed by Mohager [39]. If this model is applied to the measured elements, the evaluated creep coefficients are too large.

In order to improve model B according to Toratti [51] or the modified model B, respectively, for the evaluation of the deformation of the tests according to Mohager [39] and the deformation of the elements in Tübingen, some rheological elements for the consideration of the mechano-sorptive creep might be added to the model. However, it is doubtful, whether only the accumulated moisture should be considered. Among other influences, it seems that the rate of the moisture variation influences the creep strain as well. In order to verify this assumption tests have to be performed, where the timber elements are subjected to different climates leading to an identical moisture accumulation. Evaluating these tests with the models, an identical deformation should be determined numerically, since "only" time and accumulated moisture influence the deformation. If different creep coefficients are obtained in these tests, the rate of the moisture variation is an other parameter for the determination of the creep coefficient. Therefore this should be considered in the models by e.g. adding another set of rheological elements. However at the time being, it seems impossible to improve the models due to the missing test series.

An other approach is to develop **y**et **a**nother **r**heological **m**odel (YaRM) based on one of the explanatory models (see Sec. 2.3). From the engineering point of view, Boyd [4]'s model seems to be the most feasible explanation model, since creep and mechano-sorptive creep are explained by the reaction of the timber to forces. In this model, timber is built up of microfibrils with a viscose gel between the microfibrils (see Fig. B.1). The interlayered gel

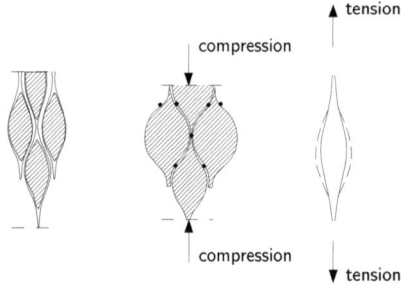

Figure B.1: Explanatory model according to Boyd [4]

is viscose and shrinks or swells, respectively. Within the model according to Boyd [4] the creep deformation is only caused by the yielding of this gel. The mechano-sorptive creep is caused by the shrinkage of the gel, leading to a gap between the gel and the microfibril, which increases the deformation of the system "microfibril & gel".

B.2 Modification of the Boyd [4]'s explanation model for the evaluation of the creep strains

In Fig. B.2 the cell-wall is simplified to a system of microfibrils and interlayered gel. For the modeling of the time-dependent behavior of timber based on the behavior of the microfibrils and interlayered gel, a subsystem can be extracted from this cell-walls (see Fig. B.3).

In order to determine the temporal behavior of this element, Boyd [4]'s model is modified in the following points and following assumptions are made, respectively.

- The gel is connected to the microfibril, so no gap between the gel and the microfibril can occur even during shrinkage. This is necessary in order to consider shrinkage in the system especially parallel to the grain.
- The gel fills the complete area between the microfibrils, so the shrinkage parallel to the grain is caused by shrinkage of the gel.
- The microfibril as well as the gel creep. If only the gel creeps, the eigenstresses due to moisture variations lead to a uniform creep deformation in the gel, resulting in elastic strains in the microfibrils. In this case due to the creeping of the gel, the deformation would decrease with time, since the eigenstresses are reduced with time.
- Shrinkage and swelling, respectively, are only caused by the gel.
- The bending stiffness of the microfibrils is infinitely large, since the compatibility and the symmetry within the system do not allow a bending of the microfibril.

B.3 Determination of the elastic properties

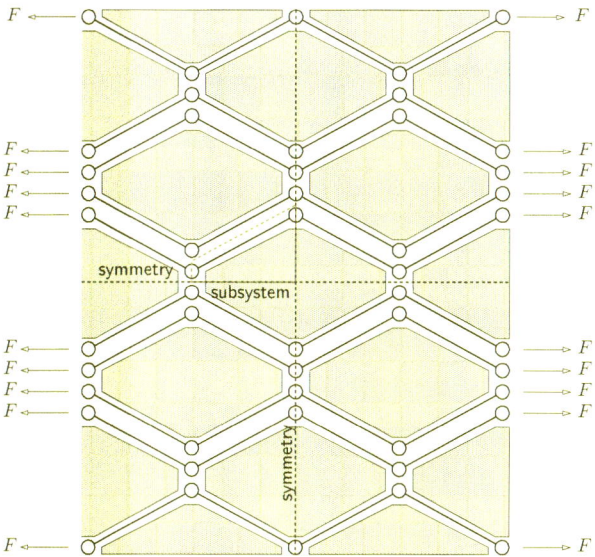

Figure B.2: Model of a part of the cell wall based on the proposal of Boyd [4]

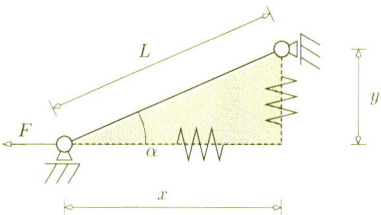

Figure B.3: Subsystem of the microfibril

- The layers S_1 and S_3 are neglected (see Fig. B.4), since in these layers a larger inclination of the microfibril related to the fibre direction of the timber exists. Therefore these layers hardly participate in the load transfer.

B.3 Determination of the elastic properties

For the determination of the stresses in the microfibril, the bedding of the microfibril on the interlayered gel is summarized to single springs. The resulting model is given in Fig. B.5. Within this structural model, it is assumed, that the Modulus of Elasticity of the gel does

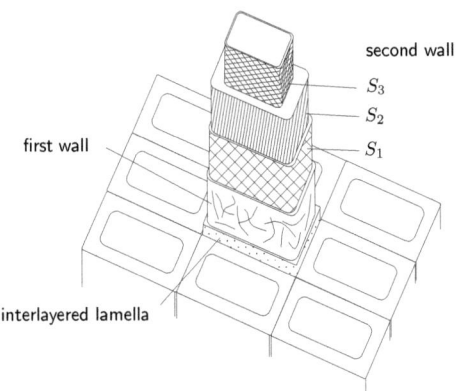

Figure B.4: Scheme of the layers S_1 to S_3 in the cell wall

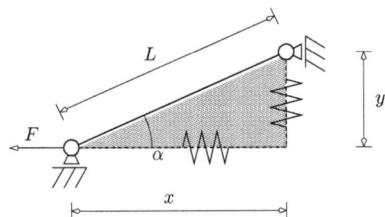

Figure B.5: Structural system of the microfibril

not depend on the directions. Therefore, the stiffness of the springs can be determined to

$$K_x = \frac{1}{2} \cdot E \cdot b \cdot \tan\alpha \text{ for } x\text{-direction} \qquad (B.1)$$

$$K_y = \frac{1}{2} \cdot \frac{E \cdot b}{\tan\alpha} \text{ for } y\text{-direction} \qquad (B.2)$$

where E Modulus of Elasticity of the interlayered gel
 b width of the fibril
 α inclination of the fibril

For the determination of the internal forces and the deformation of the system, the unknown deformations r_1 and r_2 (see Fig. B.6) are determined by the following set of equations:

$$\begin{pmatrix} K_x + \frac{EA_{fib}}{L} \cdot \cos^2\alpha & \frac{EA_{fib}}{L} \cdot \sin\alpha \cdot \cos\alpha \\ \frac{EA_{fib}}{L} \cdot \sin\alpha \cdot \cos\alpha & K_y + \frac{EA_{fib}}{L} \cdot \sin^2\alpha \end{pmatrix} \cdot \begin{pmatrix} r_x \\ r_y \end{pmatrix} = \\ \begin{pmatrix} F_x + EA_{fib} \cdot \varepsilon_{fib,inel} \cdot \cos\alpha + K_x \cdot L \cdot \varepsilon_{gel,x,inel} \cdot \cos\alpha \\ F_y + EA_{fib} \cdot \varepsilon_{fib,inel} \cdot \sin\alpha + K_y \cdot L \cdot \varepsilon_{gel,y,inel} \cdot \sin\alpha \end{pmatrix} \qquad (B.3)$$

B.3 Determination of the elastic properties

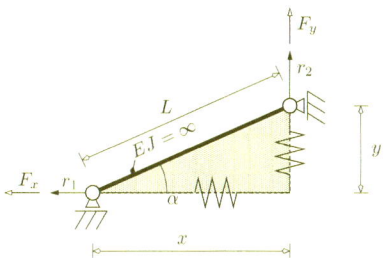

Figure B.6: Structural system for the determination of the stresses and strains of the microfibril and interlayered gel

where
- r deformation
- EA_{fib} tension stiffness of the microfibril
- $\varepsilon_{fib,inel}$ inelastic strain of the microfibril
- $\varepsilon_{gel,inel}$ inelastic strain of the interlayered gel

The strain of the gel can be determined by

$$\varepsilon_{x,tot} = \frac{r_x}{L \cdot \cos\alpha} \tag{B.4}$$

$$\varepsilon_{y,tot} = \frac{r_y}{L \cdot \sin\alpha} \tag{B.5}$$

where
- ε_{tot} total strain
- subscript x strain in x-direction (= parallel to grain)
- subscript y strain in y-direction (= perpendicular to grain)

The forces F in the subsystem can be transferred into external stresses by

$$F_x = \frac{1}{2} \cdot \sigma_x \cdot L \cdot \sin\alpha \tag{B.6}$$

$$F_y = \frac{1}{2} \cdot \sigma_y \cdot L \cdot \cos\alpha \tag{B.7}$$

where σ stress in the cell wall

Since only the cell walls transfer the forces, a vapor volume of 70% is assumed (see Fig. B.7(a)). Besides that, it is assumed, that 90% of cell-wall consists of the layer S_2. Therefore an effective volume of the layer S_2 of 27% is assumed. Due to the anisotropy of timber, the area for the load transfer parallel to the grain differs from the area perpendicular to the grain (see Fig. B.7(b)). These areas can be approximated for a cube with an edge length of 1.0 by

$$V_{\text{cell-wall}} = A_{\parallel} \cdot 1 \tag{B.8}$$

and

$$A_{\parallel} = \frac{V_{\text{cell-wall}}}{1} \tag{B.9}$$

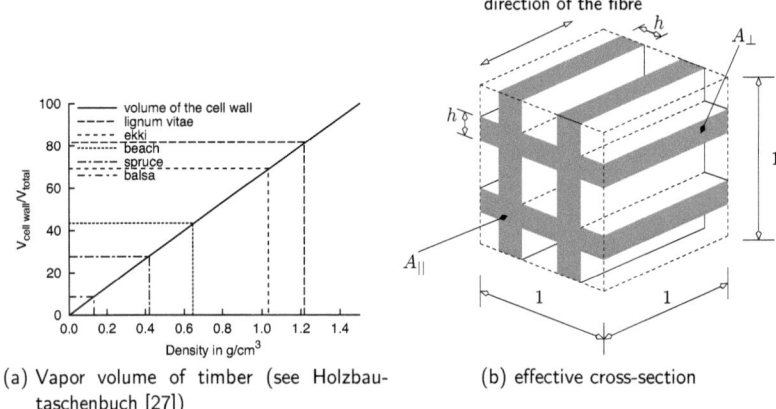

(a) Vapor volume of timber (see Holzbautaschenbuch [27])

(b) effective cross-section

Figure B.7: Vapor volume and effective cross sections

where $V_{\text{cell-wall}}$ volume of the cell-walls

For the area perpendicular to the grain $A_\perp$, it is assumed, that the area parallel to the grain $A_\parallel$ can be evaluated by

$$A_\parallel = \sum h \cdot 1 + \left(1 - \sum h\right) \cdot \sum h \tag{B.10}$$

resulting in

$$A_\perp = \sum h \cdot 1 = \left(1 - \sqrt{1 - A_\parallel}\right) \cdot 1 \tag{B.11}$$

For an assumed effective vapor volume of 0.27 the areas can be determined to be

$$A_\parallel = 0.27 \tag{B.12}$$
$$A_\perp = 0.146 \tag{B.13}$$

The vapors are considered in the determination of the stresses on the microfibril by

$$\sigma_x = \sigma_{ext} \cdot \frac{A_{tot}}{A_{\perp/\parallel}} = \sigma_{ext} \cdot \frac{1}{A_{\perp/\parallel}} \tag{B.14}$$

where σ_x stress used in Eq. (B.15) to Eq. (B.23)
  σ_{ext} nominal external stress
  $A_{\perp/\parallel}$ effective area according to Eq. (B.12) and Eq. (B.13), respectively

However, the stresses in the microfibril depend on the vapor volume, whereas the strains and the nominal stresses of the timber are independent of it, since the material properties are fitted to the global values. Therefore the vapor volume does not influence the results, as long as the parameters are consistently adapted to the boundary conditions.

Inserting Eq. (B.1) to Eq. (B.7) into each other, the following stress-strain-relation can be

B.3 Determination of the elastic properties

found for the system based on Boyd [4]'s explanation model
- parallel to grain

$$\varepsilon_{tot,x} =$$
$$\frac{2 \cdot EA_{fib} \cdot \cos^2\alpha \cdot (-1 + \cos^2\alpha)}{E \cdot (E \cdot b \cdot L \cdot \sin\alpha \cdot \cos\alpha + 2 \cdot EA_{fib} - 4 \cdot EA_{fib} \cdot \cos^2\alpha + 4 \cdot EA_{fib} \cdot \cos^4\alpha)} \cdot \sigma_y$$
$$+ \frac{E \cdot b \cdot L \cdot \sin\alpha \cdot \cos\alpha - 4 \cdot EA_{fib} \cdot \cos^2\alpha + 2 \cdot EA_{fib} \cdot \cos^4\alpha + 2 \cdot EA_{fib}}{E \cdot (E \cdot b \cdot L \cdot \sin\alpha \cdot \cos\alpha + 2 \cdot EA_{fib} - 4 \cdot EA_{fib} \cdot \cos^2\alpha + 4 \cdot EA_{fib} \cdot \cos^4\alpha)} \cdot \sigma_x$$
$$+ \frac{2 \cdot EA_{fib} \cdot \cos^2\alpha}{E \cdot b \cdot L \cdot \sin\alpha \cdot \cos\alpha + 2 \cdot EA_{fib} - 4 \cdot EA_{fib} \cdot \cos^2\alpha + 4 \cdot EA_{fib} \cdot \cos^4\alpha} \cdot \varepsilon_{fib,inel}$$
$$+ \frac{E \cdot b \cdot L \cdot \sin\alpha \cdot \cos\alpha - 4 \cdot EA_{fib} \cdot \cos^2\alpha + 2 \cdot EA_{fib} \cdot \cos^4\alpha + 2 \cdot EA_{fib}}{E \cdot b \cdot L \cdot \sin\alpha \cdot \cos\alpha + 2 \cdot EA_{fib} - 4 \cdot EA_{fib} \cdot \cos^2\alpha + 4 \cdot EA_{fib} \cdot \cos^4\alpha} \cdot \varepsilon_{gel,inel,x}$$
$$+ \frac{2 \cdot EA_{fib} \cdot \cos^2\alpha \cdot (-1 + \cos^2\alpha)}{E \cdot b \cdot L \cdot \sin\alpha \cdot \cos\alpha + 2 \cdot EA_{fib} - 4 \cdot EA_{fib} \cdot \cos^2\alpha + 4 \cdot EA_{fib} \cdot \cos^4\alpha} \cdot \varepsilon_{gel,inel,y}$$
(B.15)

- perpendicular to grain

$$\varepsilon_{tot,y} =$$
$$\frac{\cos\alpha \cdot (E \cdot b \cdot L \cdot \sin\alpha + 2 \cdot EA_{fib} \cdot \cos^3\alpha)}{E \cdot (E \cdot b \cdot L \cdot \sin\alpha \cdot \cos\alpha + 2 \cdot EA_{fib} - 4 \cdot EA_{fib} \cdot \cos^2\alpha + 4 \cdot EA_{fib} \cdot \cos^4\alpha)} \cdot \sigma_y$$
$$+ \frac{2 \cdot EA_{fib} \cdot \cos^2\alpha \cdot (-1 + \cos^2\alpha)}{E \cdot (E \cdot b \cdot L \cdot \sin\alpha \cdot \cos\alpha + 2 \cdot EA_{fib} - 4 \cdot EA_{fib} \cdot \cos^2\alpha + 4 \cdot EA_{fib} \cdot \cos^4\alpha)} \cdot \sigma_x$$
$$- \frac{2 \cdot EA_{fib} \cdot (-1 + \cos^2\alpha)}{E \cdot b \cdot L \cdot \sin\alpha \cdot \cos\alpha + 2 \cdot EA_{fib} - 4 \cdot EA_{fib} \cdot \cos^2\alpha + 4 \cdot EA_{fib} \cdot \cos^4\alpha} \cdot \varepsilon_{fib,inel}$$
$$+ \frac{2 \cdot EA_{fib} \cdot \cos^2\alpha \cdot (-1 + \cos^2\alpha)}{E \cdot b \cdot L \cdot \sin\alpha \cdot \cos\alpha + 2 \cdot EA_{fib} - 4 \cdot EA_{fib} \cdot \cos^2\alpha + 4 \cdot EA_{fib} \cdot \cos^4\alpha} \cdot \varepsilon_{gel,inel,x}$$
$$+ \frac{\cos\alpha \cdot (E \cdot b \cdot L \cdot \sin\alpha + 2 \cdot EA_{fib} \cdot \cos^3\alpha)}{E \cdot b \cdot L \cdot \sin\alpha \cdot \cos\alpha + 2 \cdot EA_{fib} - 4 \cdot EA_{fib} \cdot \cos^2\alpha + 4 \cdot EA_{fib} \cdot \cos^4\alpha} \cdot \varepsilon_{gel,inel,y}$$
(B.16)

- inelastic strains: The inelastic strains are composed of the creep strain and – for the gel – of the shrinkage and swelling, respectively

$$\varepsilon_{gel,inel,x} = \varepsilon_{gel,cr,x} + \varepsilon_s \tag{B.17}$$
$$\varepsilon_{gel,inel,y} = \varepsilon_{gel,cr,y} + \varepsilon_s \tag{B.18}$$
$$\varepsilon_{fib,inel} = \varepsilon_{fib,cr} \tag{B.19}$$

where $\varepsilon_{gel,cr,x}$ creep strain of the gel in x-direction
$\varepsilon_{gel,cr,y}$ creep strain of the gel in y-direction
ε_s shrinkage and swelling, respectively, according to Eq. (B.20)
$\varepsilon_{fib,cr}$ creep strain of the microfibril

The strain due to shrinkage or swelling is determined by

$$\varepsilon_s = \alpha_{gel} \cdot \Delta u \tag{B.20}$$

where α_{gel} shrinkage coefficient of the gel

The deformation of the model depends on five unknown parameters for the elastic deformation (see Eq. (B.15) and Eq. (B.16), respectively.)

- length of the system L
- width of the system b
- inclination of the microfibril α
- tension stiffness of the microfibril EA_{fib}
- Modulus of Elasticity of the gel E_{gel}

For the determination of these five parameters the following three boundary conditions are known:

- Modulus of Elasticity

$$E_{\text{global}} = \frac{\sigma_x}{\varepsilon_x}$$
$$= \frac{E \cdot \left(E \cdot b \cdot L \cdot sin\alpha \cdot cos\alpha + 2 \cdot EA_{fib} - 4 \cdot EA_{fib} \cdot cos^2\alpha + 4 \cdot EA_{fib} \cdot cos^4\alpha\right)}{E \cdot b \cdot L \cdot sin\alpha \cdot cos\alpha - 4 \cdot EA_{fib} \cdot cos^2\alpha + 2 \cdot EA_{fib} \cdot cos^4\alpha + 2 \cdot EA_{fib}} \tag{B.21}$$

where E_{global} global Modulus of Elasticity according to Neuhaus [43] (see Tab. B.2)
σ external stresses
ε strain according to Eq. (B.15)

As boundary condition for the evaluation of the material parameters of YaRM, the Modulus of Elasticity of timber is taken from Neuhaus [43] (see Tab. B.2) for a moisture content of 12%.

- Poisson's ratio

$$\nu = -\frac{\varepsilon_\perp}{\varepsilon_\parallel}$$
$$= -\frac{2 \cdot EA_{fib} \cdot cos^2\alpha \cdot (-1 + cos^2\alpha)}{E \cdot b \cdot L \cdot sin\alpha \cdot cos\alpha - 4 \cdot EA_{fib} \cdot cos^2\alpha + 2 \cdot EA_{fib} \cdot cos^4\alpha + 2 \cdot EA_{fib}} \tag{B.22}$$

where ν external Poisson's ratio according to Neuhaus [43] (see Tab. B.2)
$\varepsilon_\perp$ strain of the model perpendicular to the grain according to Eq. (B.16)
$\varepsilon_\parallel$ strain of the model parallel to the grain according to Eq. (B.15)

The Poisson's ratio is also taken from Neuhaus [43] (see Tab. B.2) for a moisture content of 12%.

B.3 Determination of the elastic properties

- ratio between swelling perpendicular and parallel to the grain direction

$$\frac{\alpha_\perp}{\alpha_\|} = \frac{\varepsilon_\perp(\varepsilon_{\text{shrinkage,gel}} = 1)}{\varepsilon_\|(\varepsilon_{\text{shrinkage,gel}} = 1)}$$
$$= \frac{\cos\alpha \cdot (-2 \cdot EA_{fib} \cdot \cos\alpha + 4 \cdot EA_{fib} \cdot \cos^3\alpha + E \cdot b \cdot L \cdot \sin\alpha)}{E \cdot b \cdot L \cdot \sin\alpha \cdot \cos\alpha - 6 \cdot EA_{fib} \cdot \cos^2\alpha + 4 \cdot EA_{fib} \cdot \cos^4\alpha + 2 \cdot EA_{fib}}$$
(B.23)

where $\varepsilon_\perp$ strain of the model perpendicular to the grain according to Eq. (B.16)
 $\varepsilon_\|$ strain of the model parallel to the grain according to Eq. (B.15)

For the evaluation of the parameters of the model, the following shrinkage coefficients are used:

– parallel to the grain according to Toratti [51]

$$\alpha_\| = 0.00625\%/\%\Delta u \qquad (B.24)$$

– perpendicular to the grain according to Wu and Milota [55]

$$\alpha_\perp = 0.295\%/\%\Delta u \qquad (B.25)$$

Table B.2: Modulus of Elasticity and Poisson's ratio in dependence on the moisture content for spruce according to Neuhaus [43]

MC[a]	MoE[b]	Poisson's ratio	MC[a]	MoE[b]	Poisson's ratio
0.01	12788	0.136	0.15	11614	0.256
0.02	12788	0.135	0.16	11494	0.263
0.03	12771	0.138	0.17	11364	0.269
0.04	12723	0.143	0.18	11249	0.273
0.05	12674	0.151	0.19	11123	0.276
0.06	12610	0.160	0.20	11001	0.278
0.07	12531	0.170	0.21	10893	0.279
0.08	12438	0.182	0.22	10787	0.278
0.09	12346	0.194	0.23	10684	0.277
0.10	12225	0.206	0.24	10582	0.274
0.11	12121	0.217	0.25	10493	0.269
0.12	11990	0.228	0.26	10417	0.264
0.13	11876	0.238	0.27	10341	0.257
0.14	11751	0.248	0.28	10277	0.249

[a] Moisture content
[b] Modulus of Elasticity

With these relations between the global and the local parameters, the parameters of the models can be determined. However, there are only three equations, but five unknown parameters, so two may be chosen. In this model, the geometric values L and b are chosen to

- geometric values

$$L = 1mm \tag{B.26}$$
$$b = 1mm \tag{B.27}$$

Based on these equations the elastic local material properties can be determined for a moisture content of 12%.
- inclination of the fibril

$$\alpha = 4.5° \tag{B.28}$$

- tension/compression stiffness of the microfibril

$$EA_{fib}(u = 12\%) = 1740N \tag{B.29}$$

- Modulus of Elasticity of the interlayered gel

$$E_{gel}(u = 12\%) = 1211N/mm^2 \tag{B.30}$$

B.4 Shrinkage coefficient of the gel

For the consideration of the shrinkage/swelling, it is assumed, that a shrinkage coefficient independent of the moisture content can be used as e.g. in the model according to Toratti [51]. Comparing the strain perpendicular to the grain according to Eq. (B.16) for $\Delta u = 1\%$ with the external value $\alpha_\perp = 0.295\%/(\%\Delta u)$ according to Wu and Milota [55], the shrinkage coefficient $\alpha'_{s,gel}$ is determined to

$$\alpha'_{s,gel} = 0.2968\%/(\%\Delta u) \tag{B.31}$$

In order to consider the dependence of the shrinkage/swelling on the mechanical strain, the shrinkage/swelling coefficient is modified according to Hanhijärvi [22]

$$\alpha_{s,gel} = \alpha'_{s,gel} \cdot \begin{pmatrix} 1 - 180 \cdot \varepsilon_m & \text{for } \varepsilon_m \leq 0 \\ e^{-180 \cdot \varepsilon_m} & \text{for } \varepsilon_m > 0 \end{pmatrix} \tag{B.32}$$

where $\alpha_{s,gel}$ shrinkage/swelling coefficient
$\alpha_{s,gel}$ basic shrinkage/swelling coefficient according to Eq. (B.31)
ε_m mechanical strain

B.5 Consideration of the moisture content on the elastic properties

In order to consider the influence of the moisture content on the elastic properties, the input values for the bionic approach E_{gel} and EA_{fib} are set into relation to the external parameters $E(u)$ and $\nu(u)$ according to Neuhaus [43] (see Tab. B.2). Finally equations are adapted to these values.

B.6 Consideration of creep deformations in the model

- dependence of the tension/compression stiffness of the microfibril on the moisture content based on the relations given in Neuhaus [43] (see Fig. B.8(a))

$$EA_{fib} = \Big(72667.42 \cdot u^4 - 7146.92 \cdot u^3 - 9434.89 \cdot u^2 \\ + 548.19 \cdot u + 1810.76\Big) \cdot \frac{E(u=12\%)}{11990.40} \quad \text{(B.33)}$$

where EA_{fib} tension/compression stiffness of the microfibril
u absolute moisture content
$E(u=12\%)$ Modulus of Elasticity at a moisture content of 12% in N/mm^2

Since Neuhaus [43] "only" provides values for spruce with a Modulus of Elasticity of 11990.4N/mm^2 at a moisture content of 12%, it is assumed, that the influence of the moisture content on the effective Modulus of Elasticity is affine to the values given in Neuhaus [43].

- dependence of the Modulus of Elasticity of the interlayered gel on the moisture content (see Fig. B.8(b))

$$E_{gel} = \Big(-238191.34 \cdot u^4 + 158514.23 \cdot u^3 \\ -6128.80 \cdot u^2 - 10469.53 \cdot u + 2356.34\Big) \cdot \frac{E(u=12\%)}{11990.40} \quad \text{(B.34)}$$

where E_{gel} Modulus of Elasticity of the interlayered gel
u moisture content
$E(u=12\%)$ Modulus of Elasticity at a moisture content of 12% in N/mm^2

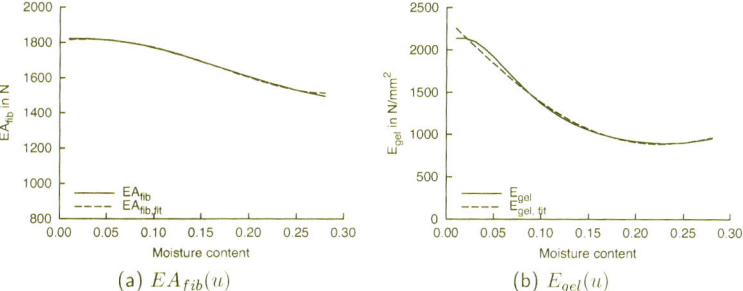

(a) $EA_{fib}(u)$ (b) $E_{gel}(u)$

Figure B.8: Material properties in dependence on the moisture content

B.6 Consideration of creep deformations in the model

B.6.1 General

In order to determine creep deformations, it is assumed that the gel, as well as the microfibrils – in difference to the proposal of Boyd [4] – are creeping. This is necessary, because only elastic deformations of the fibrils would occur if only the gel creeps. In this case, the swelling of the gel would lead to an increase of the deformation. In the course of time, the gel would creep, resulting in reduced eigenstresses. The elastic deformation of the fibril would be reduced. Since the elastic strain of the fibril is reduced, the total strain would also be reduced. So the strain of a timber element loaded by tension would be increased during wetting. In the course of time, the strain would be reduced. This reduction of the strain contradicts to the common experiences. Therefore, the gel as well as the microfibril are assumed to creep.

B.6.2 Determination of the creep parameters

B.6.2.1 Basic creep function

In this model, the mechano-sorptive creep is caused by eigenstresses in the microfibril and interlayered gel. So the dependence of the creep strain on the stress level influences the resulting deformation.

Therefore, it is assumed, that the temporal stress-strain relation of timber parallel to the grain can be described by

$$\varepsilon_{cr}(t) = \frac{\sigma}{E} \cdot \cosh(m \cdot \sigma) \cdot a \cdot t^b \qquad (B.35)$$

where $\varepsilon_{cr}(t)$ temporal creep strain
σ stress
E stiffness
t time
m,a,b parameter

Since the evaluation of the stress is split in single time steps, the dependence of the creep strain on the stress level is considered by an effective stress:

$$\begin{aligned}\Delta\varepsilon_{cr} &= \varepsilon_{cr}(\sigma + \Delta\sigma) - \varepsilon_{cr}(\sigma) \\ &= \underbrace{((\cosh(m \cdot (\sigma + \Delta\sigma)) - \cosh(m \cdot \sigma)) \cdot \sigma + \cosh(m \cdot (\sigma + \Delta\sigma)) \cdot \Delta\sigma)}_{\Delta\sigma_{eff}} \cdot \frac{a}{E} \cdot t^b\end{aligned}$$
(B.36)

where m,a,b parameter of the creep function
σ total stress of the former time steps
$\Delta\sigma$ increase of the stress within the current time step
E stiffness

Neither Eq. (B.35) nor Eq. (B.36) are suitable for the numerical solution, since in this formulation the results of every time step have to be saved.

For the numerical solution the term t^b is transferred to six Kelvin-Voigt-bodies (see Fig. B.9). The parameters of these rheological models are derived by adapting the models to the original

B.6 Consideration of creep deformations in the model

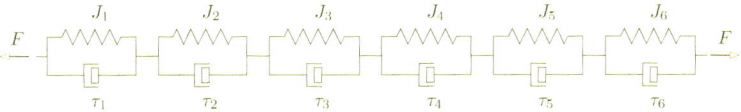

Figure B.9: Rheological model of the microfibril and the gel for the evaluation of the creep strain

equation. The increase of the creep strain within a time step $\Delta \varepsilon_{cr}$ can be evaluated by

$$\Delta \varepsilon_{cr} = \sum_{i=1}^{6} \frac{\sigma_{sum,i}}{E_{ref}} \cdot J_i \cdot \left(e^{\frac{\Delta t}{\tau_i}} - 1 \right) \tag{B.37}$$

where Δt time step
       $\Delta \varepsilon$ increase of the creep strain within a time step Δt
       $\sigma_{sum,i}$ accumulated stress of the Kelvin-Voigt-body i (see Eq. (B.38))
       E_{ref} reference MoE
       J_i, τ_i parameters of the model

The accumulated stress σ_{sum} can be evaluated by

$$\sigma_{sum,i}(t) = (\sigma_{sum,i}(t - \Delta t) + \Delta \sigma(t)) \cdot e^{-\frac{\Delta t}{\tau_i}} \tag{B.38}$$

where $\sigma_{sum,i}$ accumulated stress of the Kelvin-Voigt-body i
       $\Delta \sigma$ increase of the stress within the time step Δt
       τ_i parameter of the Kelvin-Voigt-body i

For the application of this model, the parameters of the Kelvin-Voigt-bodies τ and J_i have to be determined.

B.6.2.2 Creep parameter based on the creep behavior parallel and perpendicular to grain

In a first step, the creep parameters of the interlayered gel and the microfibrils are determined by using the creep behavior of timber parallel according to Hanhijärvi [22] and perpendicular to the grain according to Wu and Milota [55] and Hanhijärvi [23], respectively, as basic equations.

As can be seen in Fig. B.10, the creep strain in constant climate can be evaluated by the bionic approach. However, this just shows, that the fitting of the values is successful.

If the surrounding conditions vary, the creep coefficients taken from Hanhijärvi [22], Hanhijärvi [23] and Wu and Milota [55] evaluated by the bionic approach differ significantly. One reason for these differences is, that the gel creeps too much. So the eigenstresses in the gel due to swelling/shrinkage are reduced too fast by the creeping of the gel. So only little creep strains of the fibril exist.

If the failure of dry spruce perpendicular to the grain is regarded (see Fig. B.11), the weakest component is the lignin between the cells. Therefore, it is likely, that the lignin between the cells also influences the creep strain. In this case the creep coefficient perpendicular to the grain depends at least on the creep behavior of the lignin and of the microfibril. If

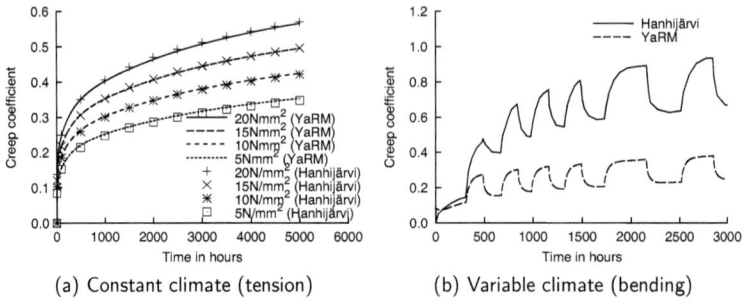

(a) Constant climate (tension)   (b) Variable climate (bending)

Figure B.10: Comparison between the bionic approach fitted to the creep coefficients and the results from literature

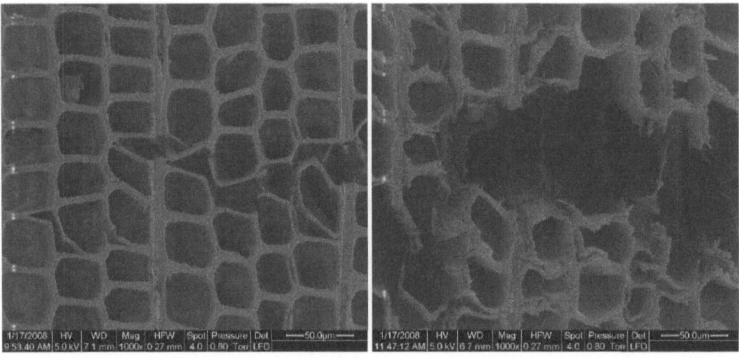

Figure B.11: Failure of dry spruce perpendicular to the grain (taken from Tukiainen and Hughes [52])

microcracks also appear during time, the temporal strain is increased additionally. Therefore, the measured creep coefficient perpendicular to the grain as e.g. given in Wu and Milota [55] or Hanhijärvi [23] does not necessarily represent the creep strain of the microfibril perpendicular to the grain. Therefore the creep parameters of the interlayered gel cannot be identified by creep tests loaded perpendicular to the grain.

B.6.2.3 Determination of the creep parameters on a simplified model

The numerical solution of the creep deformation using *kriHo* is quite sensitive concerning the input values. So the determination of the creep parameters using *kriHo* within a Newton-iteration failed, since during the interation process instabilities of the numerical solution appeared. For this reason, the model is simplified to a two layer model (see Fig. B.12).

In order to transfer the results of the simplified model into a more generalized model for

B.6 Consideration of creep deformations in the model

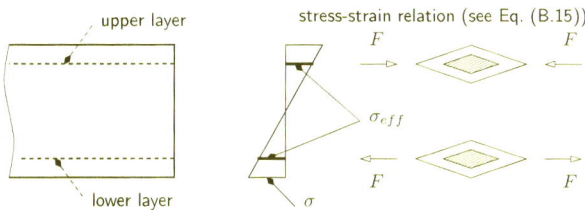

Figure B.12: Simplified model

kriHo an effective stress is used. This effective stress is approximated by following procedure, assuming a constant bending moment over time:

$$
\begin{aligned}
M &= \int \sigma \cdot z \cdot b \; dz = \int E \cdot \varepsilon_{total} \cdot z \cdot b \; dz - \int E \cdot \varepsilon_{cr} \cdot z \cdot b \; dz \\
&= \int E \cdot J \cdot \kappa - \int E \cdot b \cdot \kappa_{el} \cdot z^2 \cdot \cosh\left(m \cdot \sigma(z)\right) \cdot a \cdot t^b \; dz \equiv \kappa_{el} \cdot E \cdot J
\end{aligned}
\tag{B.39}
$$

where M bending moment
          σ stress
          E Modulus of Elasticity
          ε_{total} elastic and creep strain
          ε_{cr} creep strain
          κ_{el} elastic curvature
          h height of the cross section
          b breadth of the cross section
          z location in the cross section
and $\sigma^\star$ effective stress

The effective stress can be approximated by solving following equation:

$$
J \cdot \cosh\left(m \cdot \sigma^\star\right) = \int_{-h/2}^{h/2} z^2 \cdot \cosh\left(m \cdot \sigma_{edge} \cdot \frac{2 \cdot z}{h}\right) \; dz
\tag{B.40}
$$

where J second moment of area
          m parameter of the creep function
          h height of the cross section
          z location in the cross section
          σ_{edge} stress at the edge of the cross section
          $\sigma^\star$ effective stress

The parameters of the creep behavior of the interlayered gel and the microfibril are determined by approximating the normal creep strain according to Hanhijärvi [22]. Additionally the simplified model is adapted to the tests performed by Toratti [51] (see Fig. B.13).

In Tab. B.3 the parameters of the Kelvin-Voigt-bodies are given for the evaluation of the creep strain according to Eq. (B.37).

The creep coefficients of the components is given in Fig. B.14. As can be seen, the creep deformation under compression increases stronger than under tension. One reason for this

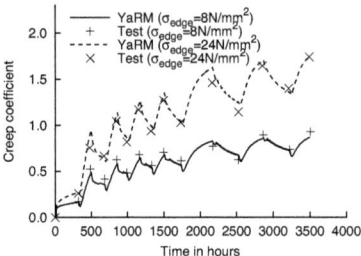

Figure B.13: Comparison between measured creep deformations by Toratti [51] (taken from Hanhijärvi [22]) and the simplified model

Table B.3: Possible parameters for the serial Kelvin-Voigt-bodies for the re-evaluation of the tests according to Toratti [51] (see Hanhijärvi [22])

	1	2	3	4	5	6
microfibril in compression						
m			-1.1367			
a			0.0039			
τ	1.6494	2	28.9002	161.7795	2360.5797	142983.70
J	1.3541	0.2121	2.953	5.4278	41.4298	401.5646
E_{ref}			$EA_{fib}(u=12\%)$			
microfibril in tension						
m			0.4684			
a			0.1333			
τ	0.1	2.9063	24.4397	218.5613	2621.4125	119998.25
J	0.9496	0.1678	0.1984	0.215	0.4632	0.6465
E_{ref}			$EA_{fib}(u=12\%)$			
interlayered gel in compression						
m			0			
a			0.5008			
τ	0.1	2	21.2317	190.9398	3177.9124	140771.25
J	0.5042	0.9119	1.8659	5.0572	20.1423	125.2196
E_{ref}			$E_{gel}(u=12\%)$			
interlayered gel in tension						
m			0			
a			0.0002			
τ	248614.16	247959.74	252718.42	72380.47	72498.86	255687.81
J	219503.74	207682.66	289904.20	-166152.17	-152111.44	688126.35
E_{ref}			$E_{gel}(u=12\%)$			

may be the buckling of the microfibrils under compression, if the gel reduces the support of the microfibril due to creeping. If some of the microfibrils buckle, the deformation increases.

B.7 Comparison of the bionic approach YaRM 0.1 to test results from literature

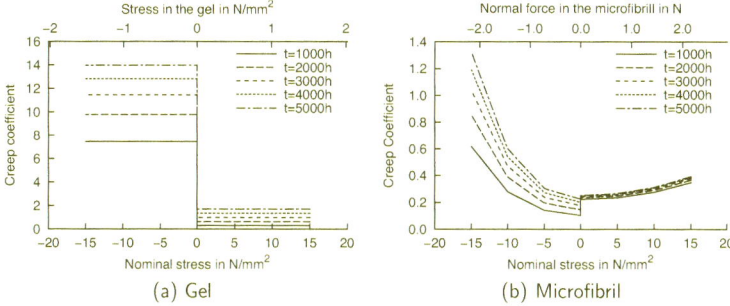

Figure B.14: Creep coefficients of the single components

Therefore the creep strain increases with time. As mentioned in Hanhijärvi [22] buckling of the microfibrils is observed by Hoffmeyer and Davidson [25], leading to the slip-plan-model for the explanation of mechano-sorptive creep in compression (see Sec. 2.3).

B.7 Comparison of the bionic approach YaRM 0.1 to test results from literature

B.7.1 Constant climate

In Fig. B.15 the resulting creep coefficients of axially loaded elements is shown. As can

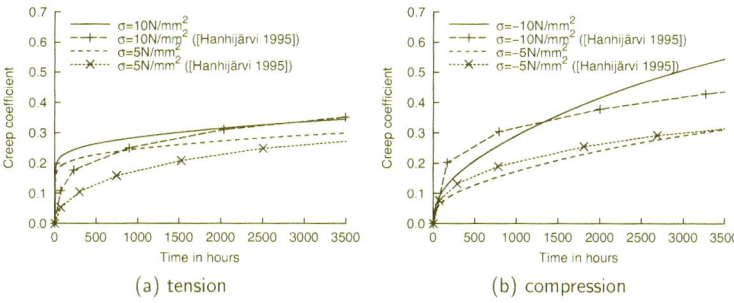

Figure B.15: Creep coefficients of axially loaded elements given by Hanhijärvi [22] and evaluated by YaRM

be seen, there are differences between the values given in Hanhijärvi [22] and the values evaluated by YaRM. The main difference appears for an element subjected to a compression of $10 N/mm^2$, whereas the other values evaluated by YaRM more or less fit the values given by Hanhijärvi [22].

B.7.2 Variable climate

In Fig. B.16 the related creep deformations of the tests according to Toratti [51] and the results of the modeling using YaRM are shown. As can be seen, there are acceptable

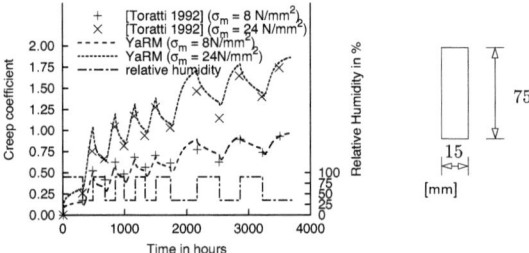

Figure B.16: Comparison between the tests according to Toratti [51] (see Hanhijärvi [22]) and the results of YaRM

differences between the tests and YaRM. However, this just shows, that the procedure of adapting the parameters of the model to the test results is successful.
In Fig. B.17 the internal stresses of the interlayered gel and the microfibril are given. As can

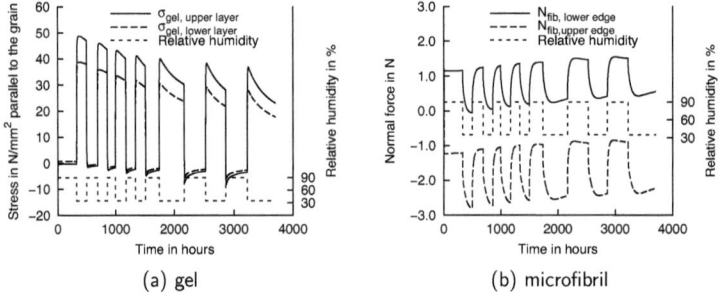

Figure B.17: Stresses in the components of the model

be seen, the moisture induced eigenstresses of the microfibril cannot be neglected. Therefore it seems logical, that these eigenstresses also lead to creep deformations, as it is assumed in this model, resulting in mechano-sorptive creep.

If moisture induced eigenstresses are one reason for the mechano-sorptive creep, the rate of the moisture variation should influence the creep coefficient, since the moisture induced eigenstresses can be reduced by creeping of the single components. Therefore the eigenstresses in a fast moisture variation should be higher than in a slow moisture variation. Due to the non-linearity of the creep deformation in dependence on the stresses the mechano-sorptive creep should be higher, the faster the moisture varies.

B.7 Comparison of the bionic approach YaRM 0.1 to test results from literature

In Fig. B.18(a) five different courses of the relative humidity are shown. All courses have in common, that the initial relative humidity is 65% and the relative humidity after 960h (= 40 days) is 40%.

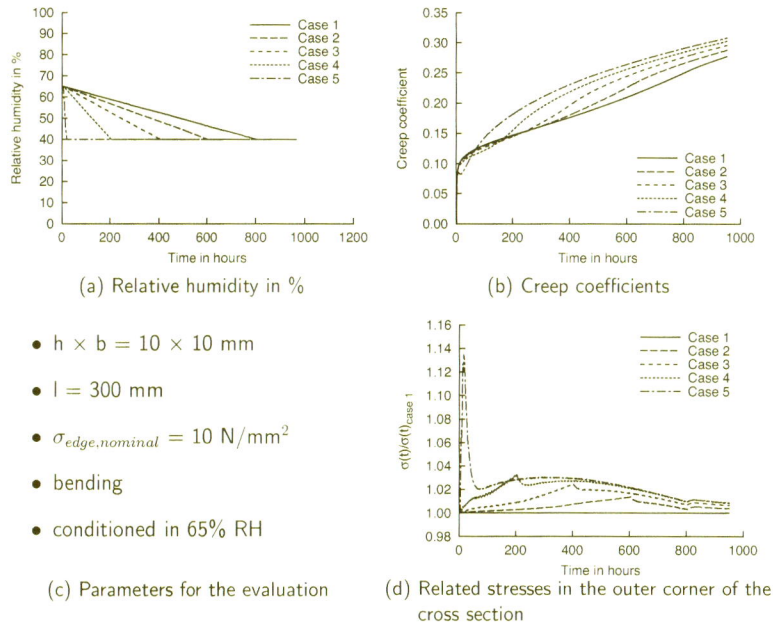

(a) Relative humidity in %

(b) Creep coefficients

- $h \times b = 10 \times 10$ mm
- $l = 300$ mm
- $\sigma_{edge,nominal} = 10$ N/mm^2
- bending
- conditioned in 65% RH

(c) Parameters for the evaluation

(d) Related stresses in the outer corner of the cross section

Figure B.18: Effects of different rates of the moisture variation on the creep coefficient

As can be seen in Fig. B.18(b), the creep coefficient increases with increasing rate of the moisture variation. Since the moisture transport is a time dependent process, the stresses in the cross section depends on the moisture distribution. So the different creep coefficient could be explained, if the stresses in these five cases differ significantly. However, as can be seen in Fig. B.18(d) the stresses between case 1 and case 2 to 4 differ only less than 3%. Due to these small differences it is very likely, that the different creep coefficients are not caused by the differences in the stresses in the outer layers due to the moisture transportation. According to this model the eigenstresses in the gel and the microfibril lead to additional creep deformations. So slow moisture variations lead to lower creep coefficients since creeping of the gel and of the fibril reduces these eigenstresses.

So the rate of moisture variation seems to be an additional parameter beside the duration of load and the accumulation of moisture, since in the evaluated examples, all five cases are subjected to the same moisture accumulation and the same duration of load.

Comparing the results of the tests by Mohager [39] with the results of the measurements (see Chap. 6), the creep coefficients in the tests were larger than the creep coefficients of the measurements, although the duration of load in the tests was shorter and the accumulated moisture of the tests was lower than in the measured elements (see Tab. B.1). However, the rate of the moisture variation in the tests was higher than in the measured elements. So the dependence of the mechano-sorptive creep on the rate of the moisture variation could explain the differences between the tests by Mohager [39] and the results of the measurements.

B.8 Conclusions

If the explanatory model according to Boyd [4] is used as basis for the modeling of the creep deformation, mechano-sorptive creep can be interpreted as the reaction of the microfibril on moisture induced eigenstresses due to the swelling or shrinkage, respectively, of the interlayered gel. So the mechano-sorptive creep is not only dependent on the moisture variation, but also on time and the rate of the changing of the moisture.

However, there are still some open questions, which should have been solved before the general application of this model.

- Hanhijärvi [22] evaluates related creep coefficients in constant climate for different stress levels. According to these results, the stress level seems to have a linear influence instead of an exponential influence as assumed in Eq. (B.35). However, the non-linearity of the creep behavior of the components influences the mechano-sorptive creep response. Therefore tests with a quite high load level should be performed in order to verify or improve, respectively, the assumed dependence of the creep deformation on the stress level.

- In *kriHo* only a 2-dimensional modeling of the creep strain is considered. However, especially shrinkage or swelling lead to a 3-dimensional loading of the microfibrils (see Fig. B.19 and Eq. (B.15)). Is it possible to neglect the influence of these 3-

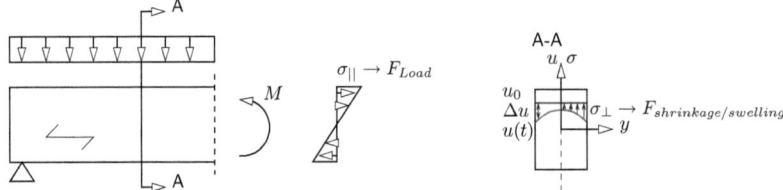

(a) stresses parallel to the grain caused by an external load

(b) stresses perpendicular to the grain caused by moisture variations Δu

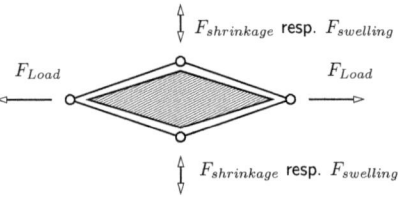

(c) resulting stresses in the microfibril

Figure B.19: Interaction of the stresses caused by shrinkage/swelling and external load

dimensional stresses? Or do these 3-dimensional stresses lead in normal cross sections

B.8 Conclusions

to deformations, which are comparable to the creep deformation of 120% of the one-dimensional stress parallel to the grain as the comparison to tests performed by Hoyle et al. [28] indicates (see Fig. B.20)?

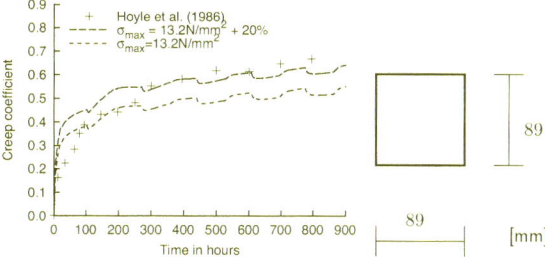

Figure B.20: Re-evaluation of the test by Hoyle et al. [28] for different stress levels

- The drying process before loading is neglected. However, during the drying process eigenstresses arise. Do they influence the time dependent behavior of timber as first evaluations by YaRM indicate (see Fig. B.21)?

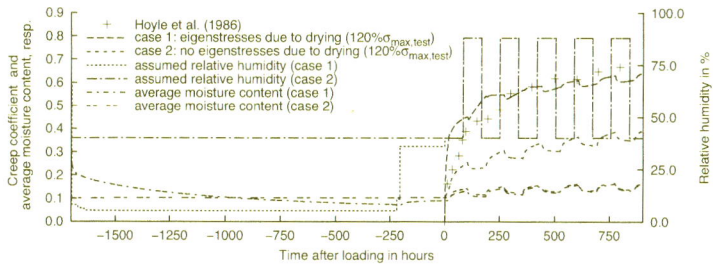

Figure B.21: Comparison of a test performed by Hoyle et al. [28] (see Toratti [51]) and evaluations by YaRM with different drying processes

- In the current modeling the angle of the microfibrils is assumed to be 0. However, there is an inclination of the microfibrils. Is there a considerable influence on the structural behavior of the fibrils?
- The layers S_1 and S_3 are neglected, since the microfibrils are orientated in larger angles than layer S_2 related to the fibre direction. Therefore, shrinkage and swelling, respectively, of the gel in the layer S_2 are hindered by these two layers (see Fig. B.22). In order to get the measured coefficient of shrinkage/swelling, the coefficient of the gel in the 3-layer model is larger than the assumed one for the model YaRM, taking only layer S_2 into account. Therefore, the eigenstresses in the 3-layer-system are larger than evaluated by YaRM.
- The parameters of the model are determined by comparing the model with some values given in Hanhijärvi [22], which cover the creep deformation for about 3500

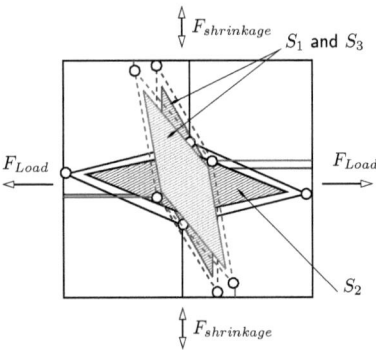

Figure B.22: Interaction of the layers S_1, S_2 and S_3

hours. Therefore it is necessary to refine the parameters of the model by comparing it with other tests, in order to use the model for the prediction beyond 3500 hours (see Fig. B.23). Otherwise larger differences in the prediction of the long term deformation

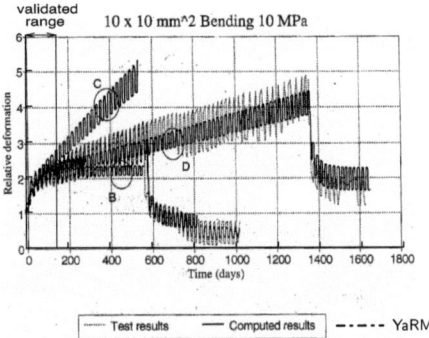

Figure B.23: Comparison between the evaluation with YaRM and the tests by Mohager [39] (taken from Toratti [51])

are expected.

C Determination of creep coefficients according to the rheological models

C.1 General

In the following sections, the creep coefficients are given, which were evaluated by the studied models according to Hanhijärvi [22], Becker [2] and Mårtensson [38]. The evaluation is performed according to the procedure described in Chap. 7. The functions are determined by adapting it to the values evaluated by *kriHo*. This curve-fitting is done by the method of least squares.

C.2 Global effects

C.2.1 Creep coefficients according to Hanhijärvi [22]

C.2.1.1 Creep coefficients for constant environmental conditions

The basic equation of the model (see Eq. (3.22)) shows, that only the change of moisture content influences the creep strain due to the dependence on $\dot{h}$. The influence of the moisture content on the Modulus of Elasticity and the influence of the Modulus of Elasticity are linear. Therefore, there is no influence of the level of the moisture content on the creep coefficient assuming constant humidity. So only the dependence on the stress level has to be taken into account.

In order to determine creep coefficients for beams, the results of the evaluations with *kriHo* are adapted to the following equation (see Fig. C.1).

$$\varphi = \frac{1}{1.102216955 - 0.014382179 \cdot |\sigma_{\max}|} \tag{C.1}$$

where $\sigma_{\max}$ maximum stress in the beam

Since the determination of the stresses is sometimes dependent on the creep coefficients and this leads to at least one iteration step, it is assumed that the maximum stress on the load level of serviceability limit state is $10 N/mm^2$, as in the former German standard DIN 1052 [8]. So Eq. (C.1) can be transformed to an equation, which depends on the ratio between external permanent load g and maximum load q. The creep coefficient after 50

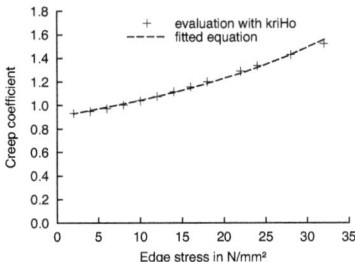

Figure C.1: Creep coefficients according to Hanhijärvi [22]

years according to Hanhijärvi [22] can be determined by

$$\varphi = \frac{1}{1.102216955 - 0.14382179 \cdot \frac{g}{q}} \tag{C.2}$$

where g permanent load
       q maximum load

C.2.1.2 Consideration of the changing moisture while reaching the equilibrium moisture content after 50 years

In order to determine the influence of the drying or of reaching the moisture equilibrium, respectively, a case study with different geometries, different stress levels and surrounding conditions is performed. The evaluated creep coefficients are compared to the fitted creep coefficient according to Hanhijärvi [22], determined in Sec. C.2.1.1. As seen in Fig. C.2, the creep coefficient is influenced by changing moisture content between ≈ -7% and ≈ 15% related to the creep coefficient in constant surrounding conditions.

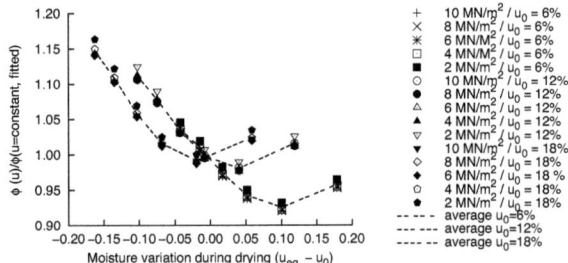

Figure C.2: Influence of the drying procedure

In Fig. C.2 no clear influence of the stress level can be determined. The average values of the different initial moisture contents are used to find an equation, which considers the drying

C.2 Global effects

process. As shown in this figure all three average curves are more or less affine. Therefore the following equation can be adapted to the results.

$$DSF(50\ years) = 6.1695 \cdot u_{eq.}^2 - 2.04657 \cdot u_{eq.} + 1.0026 + 1.85938 \cdot u_0 - 5.3990 \cdot u_0^2 \quad (C.3)$$

where $u_{eq.}$ equilibrium moisture content
       u_0 initial moisture content

With this equation, the creep coefficient based on the model according to Hanhijärvi [22] can be determined by

$$\varphi = \frac{DSF}{1.102216955 - 0.14382179 \cdot \frac{g}{q}} \quad (C.4)$$

where DSF **D**rying **s**hift **f**actor according to Eq. (C.3)
       g permanent load
       q maximum load

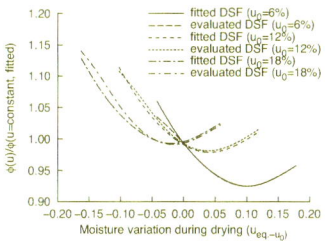

Figure C.3: Comparison between fitted and evaluated drying shift factor

C.2.1.3 Influence of the moisture content, changing in one-year sinusoidal cycles

C.2.1.3.1 Determination of the minimum moisture variation within the cross section
Since changing moisture leads to mechano-sorptive creep, the minimum moisture content has to be determined. With this minimum moisture content, the minimum mechano-sorptive creep in every point of the cross section can be evaluated. Using the equations for the moisture transportation given in Hanhijärvi [22], the following relation between the moisture variation on the surface and the minimum moisture variation in the cross section can be found: (see Fig. C.4)

- decrease of the moisture content

$$\frac{\Delta u_{min}(b)}{\Delta u_{air}} = 0.5 \cdot \tanh\left(29.686 \cdot b^2 - 26.121 \cdot b + 2.226\right) + 0.5 \leq 1.0 \quad (C.5)$$

- increase of the moisture content

$$\frac{\Delta u_{max}(b)}{\Delta u_{air}} = \begin{cases} 14.59 \cdot b^2 - 8.0517 \cdot b + 1.19 \text{ for } b < 0.30m \\ -0.269 \cdot b + 0.153 \text{ for } b > 0.30m \\ \leq 1.0 \\ \geq 0 \end{cases} \text{ for } \Delta u_{air} > 0 \quad \text{(C.6)}$$

where b thickness of the cross section in m

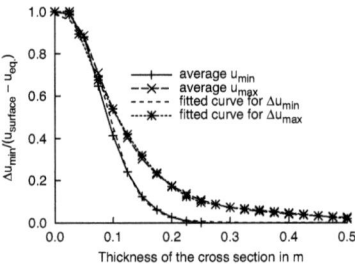

Figure C.4: Influence of the thickness on the minimal moisture variation (=moisture content in the center of the cross section)

C.2.1.3.2 Determination of creep coefficients considering the global mechano-sorptive creep For the increase of the creep coefficient due to mechano-sorptive creep the following relation can be found (see Fig. C.5)

$$GMS = \frac{\Delta u^2}{23.23077319 \cdot u_{eq.}^2 - 4.622459858 \cdot u_{eq.} + 0.4206} + 1 \quad \text{(C.7)}$$

where Δu change of the moisture content during a moisture cycle
       $u_{eq.}$ average equilibrium moisture content

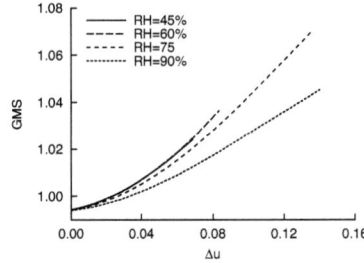

Figure C.5: Increase of the creep coefficient due to changing moisture at time t=50 years

C.2 Global effects

Therefore, the complete creep coefficient can be determined by

$$\varphi_{ms} = GMS \cdot \varphi_0 \tag{C.8}$$

where φ_0 basic creep coefficient according to Eq. (C.2)

C.2.1.4 Interaction between drying and changing moisture

For the interaction between drying and annual cycling of the moisture content, following effective moisture variation can be determined

$$\Delta u_{eff} = \sqrt{(DSF - 1) \cdot \left(23.23077319 \cdot u_{eq}^2 - 4.622459858 \cdot u_{eq} + 0.4206\right)} \tag{C.9}$$

C.2.2 Creep coefficients according to Becker [2]

C.2.2.1 Creep coefficients for constant environmental conditions

The creep coefficients according to the rheological model of Becker [2] for cross sections in constant climate can be determined to be[1] (see Fig. C.6):

$$\varphi(u_0) = m_\phi \cdot u_0 + c_\phi \tag{C.10}$$

where u_0 moisture content
m_ϕ, c_ϕ coefficients according to Tab. C.1

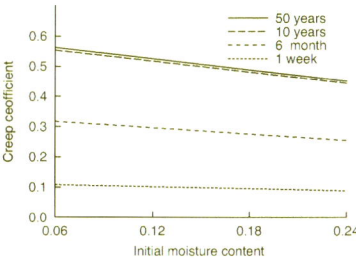

Figure C.6: Creep coefficients according to Becker [2] for constant humidity

---

[1] It has to be noticed, that in Becker [2] the creep function depends on the value "E_0". However, it is not clear, whether the Modulus of Elasticity at time $t = 0$ i.e. the first loading, or the Modulus of Elasticity with a moisture content of $u = 0$ is meant. However, for this evaluation, it is assumed that E_0 is the Modulus of Elasticity at a moisture content of $u = 0$.

In this case, the normal creep strain is independent of the moisture content. The shown influence of the moisture content on the effective creep coefficient is caused by the different elastic deformation in dependence on the moisture content.

Table C.1: Coefficients for the creep coefficients according to the model of Becker [2]

	Duration of load 50 years
m_ϕ	-0.620926538
c_ϕ	0.599995369

C.2.2.2 Consideration of the changing moisture while reaching the equilibrium moisture content after 50 years

For the consideration of the effects due to the drying, the coefficient DSF can be determined to

$$DSF(50\ years) = m_{DSF} \cdot |\Delta u| + c \qquad (C.11)$$

where m_{DSF}, c coefficients according to Tab. C.2
          Δu difference between equilibrium and initial moisture content
              $= u_{eq.} - u_0$

Table C.2: Coefficients for the drying shift factor

	m_{DSF}	c
$\Delta u > 0$	2.374552509	1.0
$\Delta u < 0$	7.655439983	1.0

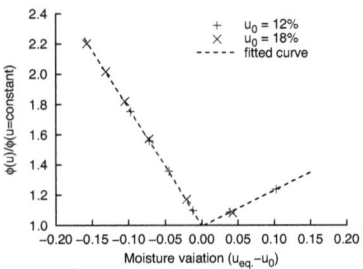

Figure C.7: DSF for the model according to Becker [2]

C.2.2.3 Influence of the moisture content, changing in one-year sinusoidal cycles

C.2.2.3.1 Determination of the minimum moisture variation within the cross section
The minimum moisture variation within a cross section can be evaluated by (see

C.2 Global effects

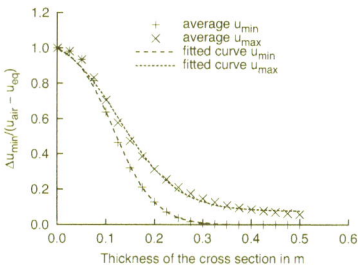

Figure C.8: Course of the annual changing moisture content in the middle of the cross section

Fig. C.8)

$$\frac{\Delta u(b)}{\Delta u_{air}} = a_{\Delta u} \cdot tanh\left(b_{\Delta u} \cdot b + c_{\Delta u}\right) + d_{\Delta u} \tag{C.12}$$

where $a_{\Delta u}, b_{\Delta u}, c_{\Delta u}, d_{\Delta u}$ coefficients according to Tab. C.3
b thickness in m

Table C.3: Coefficients for the determination of the minimum moisture variation in the cross section according to Becker [2]

Moisture variation	$a_{\Delta u}$	$b_{\Delta u}$	$c_{\Delta u}$	$d_{\Delta u}$
$\Delta u > 0$	-0.514662547	8.454162433	-1.089104367	0.590045024
$\Delta u < 0$	-0.529280327	12.20500058	-1.438451576	0.527148582

C.2.2.3.2 Determination of creep coefficients considering the global mechano-sorptive creep

The creep coefficient of Becker [2] for global mechano-sorptive creep can be determined to be (see Fig. C.9)

$$\varphi = GMS(\Delta u) \cdot \varphi_0 \tag{C.13}$$

where φ_0 basic creep coefficient according to Eq. (C.10)
$GMS(\Delta u)$ Increase due to global mechano-sorptive creep

The coefficient $GMS(\Delta u)$ can be determined to be

$$GMS = 10.19509621 \cdot \Delta u + 0.963852303 \tag{C.14}$$

C.2.2.4 Interaction between drying and changing moisture content

In order to consider the interaction between the drying and the annual cycles of the relative humidity, the drying can be transformed into an effective moisture cycle by the following

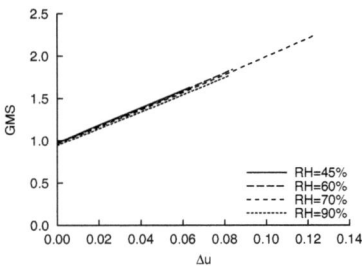

Figure C.9: Ratio of the creep coefficient in changing humidity and the creep coefficient in constant humidity

equation:

$$\Delta u_{ef} = \frac{10.19509621}{m_{DSF}} \cdot \Delta u_{\text{drying}} + 0.0035455 \tag{C.15}$$

where m_{DSF} coefficient according to Tab. C.2

C.2.3 Creep coefficients according to Mårtensson [38]

C.2.3.1 Creep coefficients for constant environmental conditions

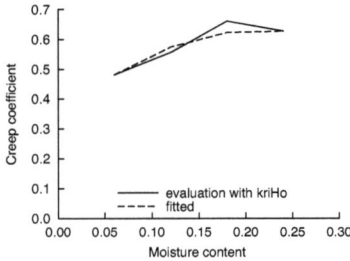

Figure C.10: Creep coefficients according to Mårtensson [38] for constant humidity

The creep coefficients according to the rheological model of Mårtensson [38] for cross sections within a constant climate for a Modulus of Elasticity of $11000 N/mm^2$ can be determined to be (see Fig. C.10):

$$\varphi(u) = a_\varphi \cdot u^2 + b_\varphi \cdot u + c_\varphi \tag{C.16}$$

where u moisture content
$a_\varphi, b_\varphi, c_\varphi$ coefficients according to Tab. C.4

C.2 Global effects

Table C.4: Coefficients for the creep coefficients according to the model of Mårtensson [38]

	Duration of load 50 years
a_φ	-6.102172594
b_φ	2.639515171
c_φ	0.345279881

C.2.3.2 Consideration of the changing moisture while reaching the equilibrium moisture content after 50 years

For the consideration of the effects due to the drying, the coefficient DSF can be determined to be

$$DSF(50 \ years) = a_{DSF} \cdot |\Delta u|^2 + b_{DSF} \cdot |\Delta u| + 1.0 \tag{C.17}$$

where $a_{DSF}, b_{DSF}, c_{DSF}$ coefficients according to Tab. C.5
  Δu difference between equilibrium and initial moisture content
  $= u_{\text{eq.}} - u_0$

Table C.5: Coefficients for the drying shift factor

	a_{DSF}	b_{DSF}
$\Delta u > 0$	0	4.9223984
$\Delta u < 0$	35.87703875	1.666899708

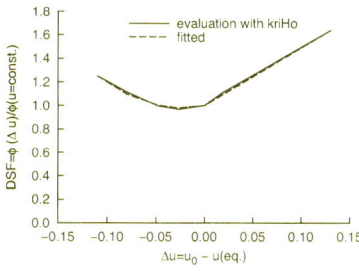

Figure C.11: Drying shift factor for the model according to Mårtensson [38]

C.2.3.3 Influence of the moisture content, changing in one-year sinusoidal cycles

C.2.3.3.1 Determination of the minimum moisture variation within the cross-section In order to determine the minimum moisture variation in the cross-section, the following equations can be fitted to the results, evaluated by *kriHo* (see Fig. C.12).

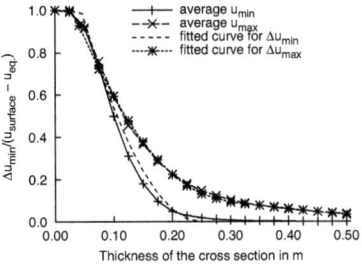

Figure C.12: Course of the annual changing moisture content in the center of the cross section

- decrease of the moisture content

$$\frac{\Delta u_{min}(b)}{\Delta u_{air}} = \left\{ \begin{array}{ll} 23.63 \cdot b^2 - 11.87 \cdot b + 1.49 & \text{for } b < 0.25m \\ \leq 1.0 & \\ \geq 0 & \\ 0 & \text{for } b > 0.25m \end{array} \right\} \text{ for } \Delta u_{air} < 0$$

(C.18)

where b thickness of the cross section in m

- increase of the moisture content

$$\frac{\Delta u_{min}(b)}{\Delta u_{air}} = \left\{ \begin{array}{ll} 12.50 \cdot b^2 - 7.44 \cdot b + 1.21 & \text{for } b < 0.30m \\ -0.30 \cdot b + 0.18 & \text{for } b > 0.30m \\ \leq 1.0 & \\ \geq 0 & \end{array} \right\} \text{ for } \Delta u_{air} > 0 \quad \text{(C.19)}$$

where b thickness of the cross section in m

C.2.3.3.2 Determination of creep coefficients considering the global mechano-sorptive creep
The creep coefficient of Mårtensson [38] for global mechano-sorptive creep can be determined to be

$$\varphi = GMS(\Delta u) \cdot \varphi_0 \tag{C.20}$$

where φ_0 basic creep coefficient according to Eq. (C.16)
$GMS(\Delta u)$ Increase due to global mechano-sorptive creep

The coefficient $GMS(\Delta u)$ can be determined to be

$$GMS = (-6.159295621 \cdot u_{eq} + 2.043398686)(100 \cdot \Delta u)^{0.55088448 \cdot u_{eq} + 0.085733826} + 1 \quad \text{(C.21)}$$

where Δu complete change of the moisture content within a cycle

C.3 Local effects

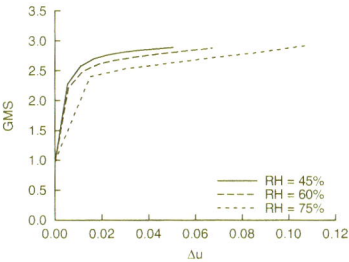

Figure C.13: Ratio of the creep coefficient in changing humidity and the creep coefficient in constant humidity

C.2.3.4 Interaction between drying and changing moisture

In order to consider the effects of drying in the determination of the creep coefficient in variable climate, the drying process is transferred to an effective moisture cycle.

$$DSF(\Delta u) \equiv GMS(\Delta u_{eff}) \tag{C.22}$$

Using this relation, the effective moisture variation due to drying can be determined to

$$\Delta u_{eff} = \frac{1}{100} \cdot {}^{0.55088448 \cdot u_{eq} + 0.085733826}\sqrt{\frac{DSF - 1}{-6.159295621 \cdot u_{eq} + 2.043398686}} \tag{C.23}$$

where Δu difference between equilibrium and initial moisture content
$= u_{\text{eq.}} - u_0$

C.3 Local effects

C.3.1 Determination of effective dimensions for the model according to Hanhijärvi [22]

In Fig. C.14 the reduction of the cross section due to the increased creep in the outer layers in dependence on the average relative humidity (see Sec. 7.3) is shown, using the model according to Hanhijärvi [22].

The reduction of the cross section can be approximated to the numerical results by following equation:

$$\Delta s = a \cdot \left(\frac{g_{perm}}{q_{total}}\right)^2 + b \cdot \left(\frac{g_{perm}}{q_{total}}\right) + c \tag{C.24}$$

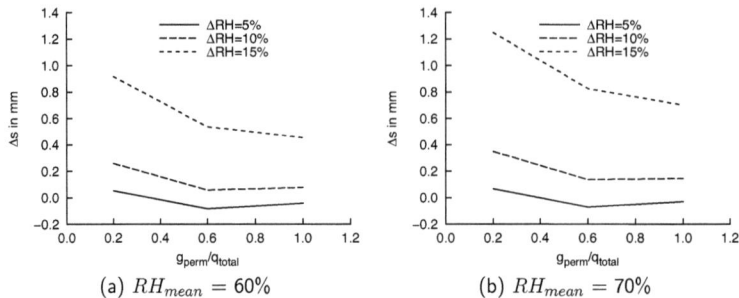

Figure C.14: Reduction of the cross section in order to consider the increased creep deformation in the outer layer based on the rheological model according to [22]

where

$$a = 0.0000325173 \cdot \Delta RH + 0.000392516 \quad (C.25)$$
$$m_b = 0.00000668077 \cdot RH_{mean} - 0.00051204 \quad (C.26)$$
$$c_b = 0.00000465472 \cdot RH_{mean} - 0.000300497 \quad (C.27)$$
$$b = m_b \cdot \Delta RH + c_b \quad (C.28)$$
$$m_c = -0.00000675262 \cdot RH_{mean} + 0.000508026 \quad (C.29)$$
$$c_c = 0.0000211535 \cdot RH_{mean} - 0.001595718 \quad (C.30)$$
$$c = m_c \cdot \Delta RH + c_c \quad (C.31)$$

Within these equations RH_{mean} indicates the average relative humidity in %, ΔRH the amplitude of the relative humidity in %, g the permanent load and q the total load.

C.3.2 Determination of effective dimensions for the model according to Becker [2]

In order to determine the influence of the changing moisture content, the reduction of the cross section dimension is also determined for the model according to Becker [2].

The reduction of the cross section dimensions can be described by the following equation (see Fig. C.16)

$$\Delta s = m_{\Delta s} \cdot \Delta RH + c_{\Delta s} \quad (C.32)$$

where ΔRH annual amplitude of the relative humidity in %
       $m_{\Delta s}, c_{\Delta s}$ coefficients according to Eq. (C.33) resp. Eq. (C.34)

$$m_{\Delta s} = 0.00000611094 \cdot RH_{mean} - 0.000194011 \quad (C.33)$$
$$c_{\Delta s} = -0.00000715701 \cdot RH_{mean} + 0.002191695 \quad (C.34)$$

C.3 Local effects

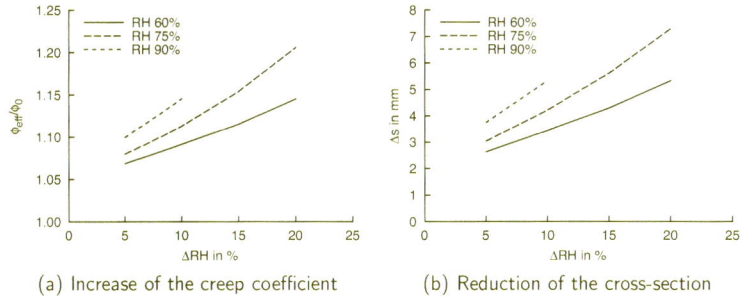

(a) Increase of the creep coefficient

(b) Reduction of the cross-section

Figure C.15: Influence of the non linear moisture distribution

where RH_{mean} average relative humidity in %

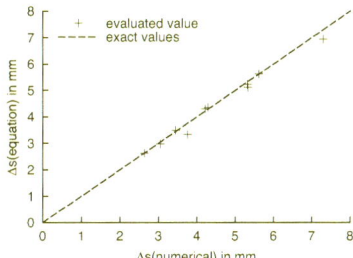

Figure C.16: Comparison of the numerically determined reduction and Eq. (C.32)

C.3.3 Determination of effective dimensions for the model according to Mårtensson [38]

In order to determine the influence of the changing moisture content in the model according to Mårtensson [38], the reduction of the cross-section dimension is determined comparable to the other models.

The reduction of the cross section can be approximated by

$$\Delta s = -1.92649 \cdot 10^{-5} \cdot \Delta RH^2 + 0.000374168 \cdot \Delta RH \qquad (C.35)$$

where ΔRH change of the relative humidity in %

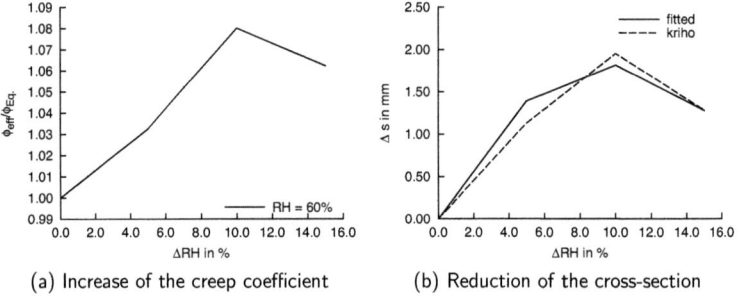

Figure C.17: Influence of the non-linear moisture distribution

D Differences in the modeling of the time-dependent behavior due to the used rheological bodies

D.1 General

For the modeling of the time-dependent behavior parallel Maxwell-bodies or serial Kelvin-Voigt-bodies are often used. For describing "only" the normal creep, both formulations are equivalent as long as the parameters fit. However in the used models, two parameters influence the time and moisture dependent behavior independently.

In the following the analytical proof should be given, why there are differences in the results dependent on the build-up of the used model.

D.2 Basic differential equation of parallel Maxwell-bodies

The temporal difference of the strain within a Maxwell-body can be described by (see Fig. D.1)

$$\frac{d\varepsilon(y)}{dy} = \frac{1}{K_i} \cdot \frac{d\sigma(y)}{dy} + \frac{\sigma(y)}{\eta_i} \tag{D.1}$$

where K_i stiffness of the spring i
   η_i damping parameter i
   y parameter (time or moisture variation, respectively)

The resulting stress is the sum of the stresses in all Maxwell-bodies. So the stress-strain relation is given by

$$\varepsilon_{res}(y) = \frac{1}{K_0 + \sum_i K_i \cdot e^{\frac{K_i}{\eta_i} \cdot y}} \cdot \sigma_{res} \tag{D.2}$$

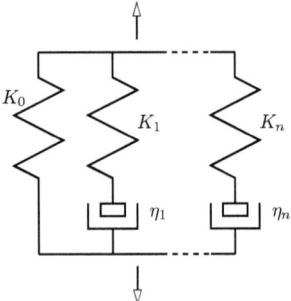

Figure D.1: Parallel Maxwell-bodies

D.3 Basic differential equation of serial Kelvin-Voigt-bodies

The incremental difference of the strain of a single Kelvin-Voigt-body (see Fig. D.2) can be determined by

$$\frac{d\varepsilon_i(y)}{dy} = \frac{1}{K_j}\frac{d\sigma_F(y)}{dy} = \frac{\sigma_D(y)}{\eta_j} \tag{D.3}$$

where K_j stiffness of the spring j
η_j damping parameter j
σ_F stress in the spring
σ_D stress in the damper

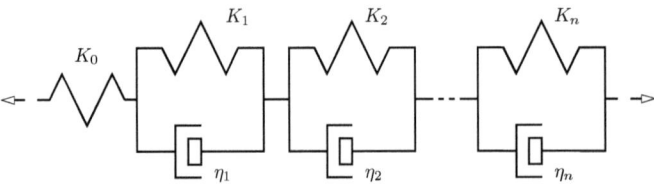

Figure D.2: Serial Kelvin-Voigt-bodies

The total strain can be evaluated in this serial model by

$$\varepsilon_{res}(t) = \sum_i \varepsilon_i \tag{D.4}$$

Therefore the stress-strain-relation is given by

$$\varepsilon_{res}(y) = \left(\frac{1}{K_0} + \sum_i \frac{1}{K_i} \cdot \left(1 - e^{\frac{-K_i}{\eta_i} \cdot y} \right) \right) \cdot \sigma_{res} \tag{D.5}$$

D.4 Equality of both principal models

Assuming, that both principal models are equal, both should lead to the same stress-strain-relation. For this reason Eq. (D.2) and Eq. (D.5) are equated, resulting in the following equation for the single parameters of the models

$$\begin{aligned}
1 = & \sum_j \frac{K_{M,0}}{K_{K,j}} \cdot \left(1 - e^{\frac{-K_{K,j}}{\eta_{K,j}} \cdot y} \right) \\
& + \frac{K_{M,0}}{K_{K,0}} + \sum_i \sum_j \frac{K_{M,i}}{K_{K,j}} \cdot e^{\frac{-K_{M,i}}{\eta_{M,i}} \cdot y} \cdot \left(1 - e^{\frac{-K_{K,j}}{\eta_{K,j}} \cdot y} \right) \\
& + \frac{1}{K_{K,0}} \cdot \sum_i K_{M,i} \cdot e^{\frac{-K_{M,i}}{\eta_{M,i}} \cdot y}
\end{aligned} \tag{D.6}$$

where K_j stiffness of the spring j
  η_j damping parameter j
  y independent parameter
 Index K property of the serial Kelvin-Voigt-body
 Index M property of the parallel Maxwell-body

This equation can be solved, as long as there is only one independent parameter. This can be done by solving the equations for different points in time, so e.g. all unknown parameters of the parallel Maxwell-bodies can be determined by the parameters of the serial Kelvin-Voigt-bodies.

However, if there are two independent parameters e.g. time and moisture variation as in the studied models, the relation between the parameters of the parallel Maxwell-bodies and the serial Kelvin-Voigt-bodies is given for $t = \infty$ by

$$1 = \left(\sum_{j=0}^{i} \frac{1}{K_{K,j}} + \frac{1}{K_{K,0}} + \sum_{j=i+1}^{n} \frac{1}{K_{K,j}} \cdot \left(1 - e^{\frac{-K_{K,j}}{\eta_{K,j}} \cdot \Delta u} \right) \right) \cdot K_{M,0} \tag{D.7}$$

where K_j stiffness of the spring j
  η_j damping parameters j
  Δu moisture variation
 Index K properties of the serial Kelvin-Voigt-bodies
 Index M properties of the parallel Maxwell-bodies

As it can be seen in Eq. (D.7) the parameters of the Maxwell-body depend on the time as well as on the moisture variation within the period of time. Therefore, the parameters of the Maxwell-bodies can not be determined irrespectivly of the moisture history in the structure. However, this contradicts the common use of the models, since the original purpose of these models is, to determine the creep deformation as a function of the parameters and not vice versa. Therefore the parameters of the parallel Maxwell-bodies cannot be determined by the parameters of the serial Kelvin-Voigt-bodies, as long as there is more than one independent parameter for the behavior of the models as for example time and moisture variation.

List of Figures

2.1	Creep phases	3
2.2	Imperfection of a column	4
2.3	Timber-concrete composite elements	4
2.4	Shed in the forest of Tübingen	5
2.5	Creep coefficients in dependence on the stress level (see Gressel [17])	6
2.6	Typical creep coefficients in dependence on the type of loading (see Gressel [17])	6
2.7	Increase of the deflection due to changing moisture content (see Hanhijärvi [22])	7
2.8	Influence of the temperature on the creep deformation after one week (see Morlier and Palka [41])	7
2.9	Boyd [4]'s model for the explanation of creep and mechano-sorptive creep	8
2.10	Creep due to the movement of the hydrogen bond (see Grossmann [19] and Hanhijärvi [22])	8
2.11	Creep as buckling of the fibrils (see Hoffmeyer and Davidson [25] and Hanhijärvi [22])	9
2.12	Reaction of the cell wall during loading (see Boyd [4] and Hanhijärvi [22])	9
2.13	Creep coefficient for solid timber in dependence on the ratio of permanent loading g to the total load p, assuming, that the live load does not affect any creep	10
2.14	Creep coefficient for solid timber in dependence on the equilibrium moisture content $u_{eq.}$ (SCL = service class)	11
2.15	Creep coefficient of solid timber in dependence on the duration of load (DOL), assuming that the complete live load consists of loads of only one class of duration of loads	12
3.1	Calculation procedure in *kriHo*	13
3.2	Model according to Toratti [51]	16
3.3	Model according to Hanhijärvi [22]	19
3.4	Rheological model according to Becker [2]	21
3.5	Influence of the thickness on the coefficient β for the evaluation of the equilibrium moisture content	24
3.6	Model according to Mårtensson [38]	25
3.7	Moisture shift factor a	26
3.8	Relation between relative humidity RH and moisture content u	30
4.1	Differences in the build-up of the models	31
4.2	Principle course of the creep strain of both typical models for normal and mechano-sorptive creep	32

4.3 Creep coefficient for constant loading, constant moisture content of 10% and an edge stress of 13.2N/mm² 33
4.4 Comparison of the creep coefficients after 50 years 34
4.5 Influence of the amplitude of the relative humidity on the creep coefficient . 35
4.6 Influence of the average relative humidity on the creep coefficient 36
4.7 Influence of the thickness of the cross section on the creep coefficient 36
4.8 Influence of the stress level on the creep coefficient 37
4.9 Influence of the Modulus of Elasticity on the creep coefficient 38
4.10 Influence of the variability of the Modulus of Elasticity on the creep coefficient according to Mårtensson [38] 38
4.11 Development of the creep coefficient of concrete in dependence on the creep coefficient of timber for constant humidity 39

5.1 Duration of tests in dependence on the number of publications (extracted from the data given in Becker [2]) 41
5.2 Comparison of the models to the test by Leivo [35] 42
5.3 Influence of the initial moisture content on the effective bending stiffness after 50 years according to Toratti [51]'s model in dependence on the amplitude of the annual changing relative humidity ΔRH (b × h = 10cm × 10cm) . . 43
5.4 Comparison of the accumulated moisture content in a cross section of h × b = 20cm × 10cm with an initial moisture content of 12% related to the same cross section with an initial moisture content of 30% in changing climate ($RH_{\text{average}} = 65\%, \Delta RH = 15\%$) 44
5.5 Ratio of the effective creep coefficient caused by mechano-sorptive creep for an initial moisture content of 12% and an initial moisture content of 30% . . 45
5.6 Influence of the imperfection on the creep coefficient 46
5.7 Definition of an accepted affine deformation 46
5.8 Course of the moisture content of a test specimen with a thickness of 2mm during the changing of the relative humidity between 30% and 90% (see Hanhijärvi [22]) 47
5.9 Maximum and minimum moisture contents of a cross section with 10cm width in dependence on the duration of the variation of the relative humidity according to the model given in Toratti [51] ($RH_{\text{average}} = 50\%, \Delta RH = 20\%$) 47
5.10 Influence of the increased creep due to daily moisture variations on the effective bending stiffness 48
5.11 Performing of the measurements 49
5.12 Roof structure of the chapel of the "Schloss Nymphenburg", Munich (see Dieringer [7]) 49
5.13 Roof structure of the Alte Münsterbauhütte, Freiburg i. Br. (see Lühr [36]) . 50
5.14 Location of the region of Tübingen (see wikipedia.de [54] and maps.google.de [37]) 51
5.15 Relative humidity in Leinfelden-Echterdingen 52
5.16 Buildings, where the measurements are performed 53
5.17 Distribution of the creep coefficient 53
5.18 Creep coefficient in dependence on the determined Modulus of Elasticity . . 54
5.19 Influences on the evaluated Modulus of Elasticity 54
5.20 Influence of the thickness of the cross section 55
5.21 Creep coefficients based on Rautenstrauch [46] (see Moorkamp [40]) 57

List of Figures

6.1 Accumulated moisture content for the elements at the Mörike-school (see Fig. 5.16) 59
6.2 Influence of the accumulated moisture in the models B and D according to Toratti [51] 60
6.3 Comparison of the measured creep deformation with the evaluation based on the model B and D according to Toratti [51] 60
6.4 Stress distribution in the tests according to Mohager [39] and stress-dependent creep coefficient as possible reasons for differences between the model B according to Toratti [51] and the tests according to Mohager [39] 61
6.5 Measured mechano-sorptive creep strain in dependence on the accumulated moisture (see Muszynski et al. [42]) 61
6.6 Modeling the mechano-sorptive creep strain based on the values given in Muszynski et al. [42] 63
6.7 Comparison of the creep strain related to the mechano-sorptive creep strain at $\sum \Delta u = 65\%$ 63
6.8 Influence of the cross section diameter on the creep coefficient 63
6.9 Model B according to Toratti [51] 64
6.10 Comparison of the modified model B according to Toratti [51] to the measured creep coefficients 65
6.11 Comparison of model B according to Toratti [51] and modified model B, respectively, with tests results (taken from Toratti [51] and added creep coefficient based on modified model B) 65
6.12 Location of Löffingen (see wikipedia.de [54] and maps.google.de [37]) . . . 66
6.13 Buildings, where the measurements where performed by Gutenkunst [20] . . 66
6.14 Surrounding conditions in Löffingen, region of Breisgau-Hochschwarzwald . . 67
6.15 Temperature in Lenzkirchen 68
6.16 Comparison between evaluated and measured creep coefficients, which are related to the elastic deformation due to permanent dead load 68
6.17 Accumulated relative humidity and accumulated snow height in the region of Tübingen and Breisgau-Hochschwarzwald 69

7.1 Principal consideration of changing moisture 72
7.2 Concept of the consideration of time-dependent effects 73
7.3 Evaluated creep coefficients by means of the model B according to Toratti [51] and the modified model B 74
7.4 Creep coefficients based on the model B according to Toratti [51] and the modified model B (see Eq. (7.5)) 75
7.5 Drying shift factor for the model according to Toratti [51] and the modified model B, respectively, according to Eq. (7.6) (for $k_{model}(DSF)$ see Eq. (7.10)) 75
7.6 Minimum moisture content in the cross section and idealized distribution of the creep coefficient due to mechano-sorptive creep 77
7.7 Assumed moisture transportation 78
7.8 Influence of the thickness on the minimal moisture variation (=moisture content in the center of the cross section) 78
7.9 Relation between change of relative humidity and equilibrium moisture content 79
7.10 Dependence of the coefficient GMS on the changing moisture content Δu for the model B according to Toratti [51] and the modified model B, respectively 80

7.11 Maximum and minimum moisture content, respectively, of a 100mm × 200mm timber beam, subjected to an annual cycle of the relative humidity with an amplitude of 10% 82
7.12 Comparison of the creep coefficients in changing humidity
φ_{kriHo} = creep coefficients considering the increased moisture variations in the outer layers
$\varphi_{eq.}$ = creep coefficients considering "only" the global mechano-sorptive creep 83
7.13 Minimum amplitude of the annual cycle of the relative humidity ΔRH_{min} for $GMS(\Delta RH_{min}) = 0.95 \cdot GMS_{max}$) 84

8.1 Different scenarios concerning the load history 85
8.2 Influence of a live load once in 50 years (cases 1 and 2) on the creep coefficient
p_{eff} = quasi-permanent part of the live load
p_{total} = total live load 86
8.3 Influence of a repeating load (case 3, e.g snow load) on the creep coefficient in constant climate. 87
8.4 Comparison between the coefficient ψ evaluated by kriHo and using Eq. (8.9) 88
8.5 Load scenarios 89
8.6 Maximum and minimum coefficient ψ, respectively, for an annually repeating load according to Eq. (8.9) 89
8.7 Climate in the region of Löffingen 90
8.8 Considered annual loading and course of relative humidity 91
8.9 Ratio of the quasi-permanent load and permanent load ψ evaluated by kriHo using the modified model B for different average relative humidities RH, different amplitudes ΔRH and different cross section dimensions b 91
8.10 Assumed average moisture content for the climate given in Fig. 8.8(a) (RH=60%) 91
8.11 Schematic explanatory model of the remaining creep strain for an annually repeating load in variable moisture contents 92
8.12 Comparison of the quasi-permanent part of the live load ψ in variable surrounding conditions according to Eq. (8.13) and evaluated by kriHo 94
8.13 Course of the relative humidity in Löffingen for the determination of the quasi-permanent part of the snow load 94

9.1 System studied for the effects of the increased creep deformations 98
9.2 Critical ratio $N_{\text{permanent}}/N_{\text{total}}$, at which creep deformations should be considered in order to achieve the same level of acceptance as proposed in DIN 1052 [9], depending on the creep coefficient and the slenderness of the column 101
9.3 Temporal development of the creep coefficient of concrete dependent on the creep coefficient of timber 103

A.1 Comparison of the creep coefficient between the model B according to Toratti [51] and kriHo 113
A.2 Comparison of the moisture content between Fragiacomo [15] and kriHo . . 114
A.3 Comparison of the creep coefficient between Hanhijärvi [22], p.113 and kriHo for constant moisture content 114
A.4 Comparison of the creep coefficient between Hanhijärvi [22], p.122 and kriHo for changing moisture content and different bending moments 115

List of Figures

A.5 Comparison of the moisture content between Hanhijärvi [22], p.120 and *kriHo* 115
A.6 Comparison of the creep coefficients due to linear creep between Becker [2] and *kriHo* 116
A.7 Comparison between Becker [2] and *kriHo* 116
A.8 Comparison between Mårtensson [38], p. 113 and *kriHo* 117

B.1 Explanatory model according to Boyd [4] 120
B.2 Model of a part of the cell wall based on the proposal of Boyd [4] 121
B.3 Subsystem of the microfibril 121
B.4 Scheme of the layers S_1 to S_3 in the cell wall 122
B.5 Structural system of the microfibril 122
B.6 Structural system for the determination of the stresses and strains of the microfibril and interlayered gel 123
B.7 Vapor volume and effective cross sections 124
B.8 Material properties in dependence on the moisture content 129
B.9 Rheological model of the microfibril and the gel for the evaluation of the creep strain 131
B.10 Comparison between the bionic approach fitted to the creep coefficients and the results from literature 132
B.11 Failure of dry spruce perpendicular to the grain (taken from Tukiainen and Hughes [52]) 132
B.12 Simplified model 133
B.13 Comparison between measured creep deformations by Toratti [51] (taken from Hanhijärvi [22]) and the simplified model 134
B.14 Creep coefficients of the single components 135
B.15 Creep coefficients of axially loaded elements given by Hanhijärvi [22] and evaluated by YaRM 135
B.16 Comparison between the tests according to Toratti [51] (see Hanhijärvi [22]) and the results of YaRM 136
B.17 Stresses in the components of the model 136
B.18 Effects of different rates of the moisture variation on the creep coefficient 137
B.19 Interaction of the stresses caused by shrinkage/swelling and external load 138
B.20 Re-evaluation of the test by Hoyle et al. [28] for different stress levels 139
B.21 Comparison of a test performed by Hoyle et al. [28] (see Toratti [51]) and evaluations by YaRM with different drying processes 139
B.22 Interaction of the layers S_1, S_2 and S_3 140
B.23 Comparison between the evaluation with YaRM and the tests by Mohager [39] (taken from Toratti [51]) 140

C.1 Creep coefficients according to Hanhijärvi [22] 142
C.2 Influence of the drying procedure 142
C.3 Comparison between fitted and evaluated drying shift factor 143
C.4 Influence of the thickness on the minimal moisture variation (=moisture content in the center of the cross section) 144
C.5 Increase of the creep coefficient due to changing moisture at time t=50 years 144
C.6 Creep coefficients according to Becker [2] for constant humidity 145
C.7 DSF for the model according to Becker [2] 146

C.8 Course of the annual changing moisture content in the middle of the cross section 147
C.9 Ratio of the creep coefficient in changing humidity and the creep coefficient in constant humidity 148
C.10 Creep coefficients according to Mårtensson [38] for constant humidity ... 148
C.11 Drying shift factor for the model according to Mårtensson [38] 149
C.12 Course of the annual changing moisture content in the center of the cross section 150
C.13 Ratio of the creep coefficient in changing humidity and the creep coefficient in constant humidity 151
C.14 Reduction of the cross section in order to consider the increased creep deformation in the outer layer based on the rheological model according to [22] 152
C.15 Influence of the non linear moisture distribution 153
C.16 Comparison of the numerically determined reduction and Eq. (C.32) 153
C.17 Influence of the non-linear moisture distribution 154

D.1 Parallel Maxwell-bodies 156
D.2 Serial Kelvin-Voigt-bodies 156

List of Tables

2.1	Range of the creep coefficient of timber k_{def} according to different standards Eurocode 5 [14], DIN 1052 [8] and DIN 1074 [11]	11
3.1	Parameters for the model according to Toratti [51]	17
3.2	Parameters for the rheological model according to Hanhijärvi [22]	19
3.3	Parameters for the model according to Becker [2]	21
3.4	Coefficients A, B and C for the determination of the limit of proportionality	23
3.5	Material parameters τ_n and J_n (creep data 2, see Mårtensson [38])	25
3.6	Intervals for the moisture shift factor a	26
3.7	Material properties for mechano-sorptive creep and recovery (see Mårtensson [38])	27
3.8	Proposed values for the shringake/swelling coefficient	28
3.9	Modulus of Elasticity for three relative humidities	29
3.10	Parameters, used in the simulations	29
4.1	Creep coefficient for constant loading, constant moisture content of 10% and an edge stress of 13.2N/mm^2	33
5.1	Used elements for the determination of the creep coefficient in the region of Tübingen	52
5.2	Average values and standard deviations of the creep coefficient of Tübingen	52
5.3	Creep coefficient according to Rautenstrauch [46] and Moorkamp [40] without elastic recovery	57
6.1	Parameters for the description of the mechano-sorptive creep used in the model according to Muszynski et al. [42]	62
6.2	Parameters	64
6.3	Measured elements	67
7.1	Parameters for the creep coefficients in constant surrounding conditions evaluated by the model B according to Toratti [51] and by the modified model B	74
7.2	Coefficients for the drying shift factor	76
8.1	Coefficients m and c_{DOL} for the determination of the quasi-permanent part of an annually repeating live load	90
8.2	System parameter	93
8.3	Comparison between the effective part of the snow load $\psi \cdot s$ in the measurements according to Gutenkunst [20] and according to Eq. (8.13)	96
9.1	Range of parameters for the determination of the critical ration $N_{\text{permanent}}/N_{\text{total}}$	100
9.2	Parameters for Eq. (9.14)	101

9.3	Modified intervals for the determination of the effective creep coefficients in timber-concrete composite according to [48] due to the increased creep coefficient of the modified model B	103
A.1	Creep coefficients for linear creep according to Becker [2]	115
B.1	Key parameters of the time dependent behaviour	119
B.2	Modulus of Elasticity and Poisson's ratio in dependence on the moisture content for spruce according to Neuhaus [43]	127
B.3	Possible parameters for the serial Kelvin-Voigt-bodies for the re-evaluation of the tests according to Toratti [51] (see Hanhijärvi [22])	134
C.1	Coefficients for the creep coefficients according to the model of Becker [2]	146
C.2	Coefficients for the drying shift factor	146
C.3	Coefficients for the determination of the minimum moisture variation in the cross section according to Becker [2]	147
C.4	Coefficients for the creep coefficients according to the model of Mårtensson [38]	149
C.5	Coefficients for the drying shift factor	149

*... den Vorhang zu
und alle Fragen offen.* [a]
*(Bertholt Brecht,
Der gute Mensch von Sezuan)*

---

[a] ... the curtain falls
and all questions remain unanswered.

I want morebooks!

Buy your books fast and straightforward online - at one of world's fastest growing online book stores! Environmentally sound due to Print-on-Demand technologies.

Buy your books online at
www.morebooks.shop

Kaufen Sie Ihre Bücher schnell und unkompliziert online – auf einer der am schnellsten wachsenden Buchhandelsplattformen weltweit! Dank Print-On-Demand umwelt- und ressourcenschonend produziert.

Bücher schneller online kaufen
www.morebooks.shop

KS OmniScriptum Publishing
Brivibas gatve 197
LV-1039 Riga, Latvia
Telefax: +371 686 204 55

info@omniscriptum.com
www.omniscriptum.com

Printed by Books on Demand GmbH, Norderstedt / Germany